EquipThink™

to achieve optimal results

The Exemplary Worker Book Series

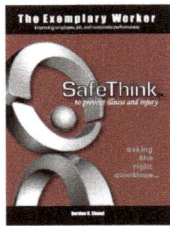

SafeThink™ ...to prevent illness and injury

SafeThink is a structured critical thinking strategy you can use to identify, predict, and control hazardous situations before, during, and after completing work. This cognitive-based safety strategy can be used on the fly, at work, at home, at play, and while driving. *SafeThink* also provides strategies for you to remain focused on your tasks.

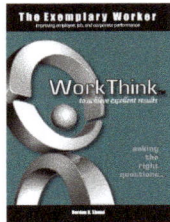

WorkThink™ ...to achieve excellent results

WorkThink is a thinking strategy you can use to achieve quality results with the least amount of effort. It usually takes little extra effort to do quality work instead of inferior work. *WorkThink* also emphasizes understanding the expectations of your supervisor, team leader, and customers so that you can achieve the excellent results they expect.

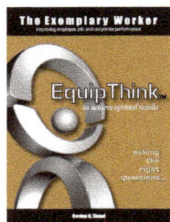

EquipThink™ ...to achieve optimal results

EquipThink is a thinking strategy for you to use tools, mobile equipment, and stationary equipment effectively and efficiently. The goals are for you to achieve the desired results with minimal stress on equipment, to conserve energy, and to extend equipment life. The input–process–output thinking strategy, in conjunction with identifying critical variables, is used to achieve optimal results.

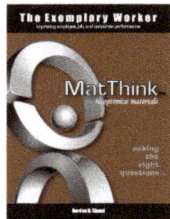

MatThink™ ...to optimize materials

MatThink is a thinking strategy you can use to make the most effective use of materials. The thinking strategy applies to recovering, processing, modifying, applying, transporting, and storing materials. Because equipment and materials are usually closely related, the input–process–output thinking strategy, in conjunction with identifying critical variables, is used to optimize material recovery and use.

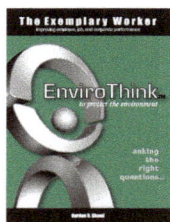

EnviroThink™ ...to protect the environment

Both industry and individuals have a responsibility to protect the environment. *EnviroThink* is a critical thinking strategy you can use to identify and respond to environmental issues for any job position that you might hold. *EnviroThink* helps you think through your work by asking yourself specific questions relating to environmental issues important to organizations.

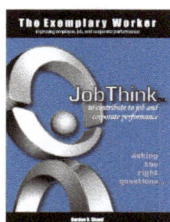

JobThink™ ...to contribute to job and corporate performance

Exemplary workers understand what is important to their organizations. They know the issues critical to business success and where to focus their efforts. *JobThink* addresses the critical thinking strategies you can use to identify what is important for job and corporate performance.

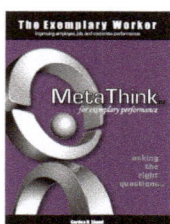

MetaThink™ ...for exemplary performance

MetaThink applies some of the thinking strategies addressed in previous books in different ways and also addresses new thinking strategies useful for the workplace. You can use these thinking strategies, along with the detailed thinking strategies addressed in other books of this series, to achieve exemplary performance.

The Exemplary Worker Book Series

"Rarely can workers from any sector access self-paced instructional materials that are easy-to-use, step-by-step guides to workplace learning. *The Exemplary Worker* book set is an exception. These books offer a good breadth of learning for workers in contexts ranging from: exemplary performance; job and corporate performance; results optimization; and work excellence. With meticulous organization, these essential training references are helpful guides for workers seeking to improve their performance. With prefaces designed to help trainers/instructors assist workplace learners, these books use critical thinking strategies that identify what matters to workers and supervisors considering people, equipment, materials, environments, and organization in concert."

—**Eugene G. Kowch, Ph.D.**, Leading Complex and Adaptive Learning
Systems/Organizations, University of Calgary, Canada

"The power of thinking in determining our safety, health, and welfare is obvious, but how to manage such cognition or self-talk for injury prevention, self-motivation, and self-improvement is not so obvious. Answers are provided in this action-focused series of self-help books on *The Exemplary Worker* by Gordon D. Shand. He offers much practical information for leadership, safety, and well-being. Each of these books provides critical and structured thinking strategies for optimizing performance on several fronts, from improving safety and productivity in the workplace to actively caring as a teacher, parent, or friend."

—**E. Scott Geller, Ph.D.**, author of The Psychology of Safety Handbook; Alumni Distinguished
Professor, Virginia Tech; Senior Partner, Safety Performance Solutions

"These are very practical books. I, myself, have been interested in the fundamental processes of human thinking. For creativity there is Lateral Thinking. For exploration there is the parallel thinking of the Six Thinking Hats. For perception there is the CoRT school programme. *The Exemplary Worker* series of books provide frameworks for focused thinking about specific situations. The frameworks guide the thinker to deal with the situation instead of messing about. That is why the books are so practical."

—**Dr. Edward de Bono**, Author of Lateral Thinking and
Six Thinking Hats and creator of CoRT

The Exemplary Worker Book Series

EquipThink™

to achieve optimal results

Gordon D. Shand

HDC Human Development Consultants Ltd.
PO Box 4710, Edmonton, AB, Canada T6E 5G5
www.hdc.ca
www.safethink.ca

EquipThink™ is a trademark of HDC Human Development Consultants Ltd.

Library and Archives Canada Cataloguing in Publication
Shand, Gordon D.
   EquipThink to achieve optimal results / Gordon D. Shand.
(The exemplary worker book series)
ISBN 978-1-55338-034-4
   1. Industrial equipment--Utilization. 2. Critical thinking.
3. Industrial equipment--Maintenance and repair. 4. Employees--
Training of. I. HDC Human Development Consultants II. Title.
III. Series: Exemplary worker
HD39.3.S52 2014   338.4'5   C2014-902765-0

Published by HDC Human Development Consultants Ltd.

Published in Canada

HDC *Human Development Consultants Ltd.*

Website: www.hdc.ca
E-mail: hdc@hdc.ca
Phone: (780) 463-3909

Acknowledgements

Developing *The Exemplary Worker* book series has been challenging and rewarding. I am certainly grateful for all the help I have received to produce quality products. Over one hundred people have contributed to the quality of the content and presentation.

Generally, I developed the first draft of the books working on evenings and weekends. I would blitz the first draft for a book—I produced the draft in a month to three months. During those times, my family's gracious support allowed me to concentrate on the task and to dialogue with them about the concepts. Once a first draft was produced, consultants in my firm carried out several edits as time allowed. HDC's Production Department developed illustrations and formats to produce a book ready for validation by industry. Because the people from industry volunteered their time and some validations were conducted in sequence, the validation process for each book took up to six months or more.

Many staff contributed to the development process. I would like to acknowledge those consultants who struggled to gather relevant content when working with customers—they gave cause to identify the thinking strategies used by exemplary workers and to develop the training for HDC consultants. Many thanks to the consultants who worked so diligently with me to produce the books. They were adamant in adhering to our standards for quality, even when I was burned out and wanted to put closure to a topic. Thanks to Janelle Beblow, Art Deane, Alice Graham, Jean MacGregor, and Bruno Schoenfelder for the wonderful edits and feedback. Thanks to Phil Jenkins, Kris Vasey, and Denise Hodgins for developing the illustrations, formatting the documents, and creating the book covers. Thanks to Maria Peck for coordinating the validations and field tests and proofing text. Their personal support, commitment to quality, and attention to detail are greatly appreciated.

I have been exceptionally fortunate to work with so many wonderful people from industry. They have been great mentors—they have made many contributions to my personal growth. A special thanks to nearly a hundred people who have volunteered their time to validate and field test the strategies.

Who is *The Exemplary Worker* series for?

The Exemplary Worker series benefits:

- **Individuals** who want to have outstanding performance

- **Apprentices and students** who want to work safely and effectively

- **Supervisors** who want staff to be more effective

- **Trainers** who want to contribute to improved corporate, job, and employee performance
- **Trades and technology instructors** who want their apprentices and students to work safely and effectively

- **Instructional designers** who want to ensure that training is relevant, useful, and practical

- **HR managers** who want to improve the development and retention of exemplary workers

- **Operations staff** who want to optimize production and minimize losses

Contents

Table of Contents (continued)

Preface

In addition to being skilled, exemplary workers use a broad range of *critical thinking strategies* to maintain outstanding performance. Exemplary workers know what is important to their jobs and organizations—they put their efforts in the right places by doing the most important things, doing them effectively, and doing them efficiently. Because they know what is important to the job and the organization, they effectively coordinate their actions with others and make decisions in the best interest of their organizations. Knowledge and thinking skills empower workers to achieve exemplary performance, be flexible as workplaces continue to evolve, and provide leadership within the workplace.

Exemplary performance can have many benefits for you, the line worker, lead operator, foreman, or supervisor, including:
- increased job satisfaction
- being recognized by your peers and supervisors as an effective employee
- increased potential for keeping your job during slow economic times
- increased potential for receiving salary/wage increases or bonuses
- increased opportunity for new or different work assignments
- increased potential for promotion

Each of the seven books in *The Exemplary Worker* series focuses on one of five domains (**PEMEO**):
- **P**eople
- **E**quipment
- **M**aterials
- **E**nvironment
- **O**rganization

Loss and/or optimization (LO) are the main themes for the domains, creating the word **LO-PEMEO™**. LO-PEMEO stands for Loss and Optimization of People, Equipment, Materials, Environment, and Organization. As an example: **L**oss to **P**eople is illness and injury; **O**ptimizing **P**eoples' performance is working effectively and efficiently; **L**oss to **E**quipment is damage and shortened operating life; and **O**ptimizing **E**quipment is using equipment effectively and efficiently. The books place a strong emphasis on using **thinking strategies** and **asking quality questions**—the goals are to minimize losses and optimize performance of PEMEO.

The series of books addresses both loss and optimization of each domain. We recommend that you complete each of the first six books in the sequence. However, the books can be studied in any order without difficulty. The last book in the sequence, *MetaThink*, should be read last. *MetaThink* applies some of the thinking strategies addressed in previous books but in different ways and also addresses new thinking strategies useful for the workplace.

Introduction to *The Exemplary Worker* Series

Over the last twenty-five years, the process of discovering *what's important* for exemplary worker performance has gone full circle. The process began for me when I interviewed exemplary workers to identify relevant training content. My premise was that exemplary workers know what is important for people to do their jobs effectively. Over time, it became apparent to me that one of the reasons exemplary workers perform so well is that they use a set of generic thinking strategies. After starting a consulting firm to design and develop training, I developed a comprehensive internal training program for our consultants and technical writers who develop training programs. The training focused on using generic thinking strategies and critical questions to identify training content that helps workers perform effectively. With a lot of support, I have revised our consultant training program and made it available to the public for people to learn and refine their personal thinking strategies to be exemplary workers.

The Exemplary Worker books are presented as a series. The same concepts underlie all seven books. For example, a safety incident may cause harm to a person and result in other losses—work may be suspended, equipment and materials damaged, the environment harmed. The organization could also experience unpredicted costs and have its reputation harmed. This introduction provides a framework and the key concepts that apply to the series. The discovery process and happenstances that led to the development of *The Exemplary Worker* series are explained to provide a setting and context to give meaning to the underlying concepts.

The Discovery Process

For me, the real discovery process began in 1985 when I founded the consulting firm HDC Human Development Consultants Ltd. (HDC) to design and develop customized technical training programs. I believed that it was possible to develop quality training for any industry without having an in-depth understanding of the organization, its technology, or the tasks that its people perform. The premise was that a well-thought-out instructional design and development process combined with effective consulting skills would be sufficient.

As founder of the company, I felt that I was successful in providing leadership to identify training content important to my customers—customers often asked me to do additional work. If I could do the work well, then certainly others in the firm could as well and, for some deliverables, do better.

The Plan

The plan was that I would work with customers to develop the outline of the training program (curriculum) and identify critical content for the program. The training program would be documented in one of three ways:
- a list of specific courses
- a list of general training objectives
- a competency-based training profile

Competency-Based Training Profile

The following illustration is a *partial* example of a competency-based training profile. The profile is a visual presentation of the competencies (tasks and support knowledge) that specific work groups require to do their work safely and effectively.

ORIENTATION	Complete Company Orientation	Describe Roles and Responsibilities	Identify Local Structures and Facilities	Describe and Use Communication Systems	Identify Customers and their Expectations
SAFETY	Describe and Use Personal Safety Equipment	Review Safety Handbook	Complete First Aid Training	Decribe and Operate Personal Gas Monitors	Describe Codes of Practice
ENVIRONMENT	Describe Environmental Responsibilities	Describe and Store Hazardous Wastes	Describe and Monitor Gas Emissions	Take Waste Water Samples	Describe and Participate in Spill Response Exercises
GENERAL KNOWLEDGE AND SKILLS	Describe Flammable Gas Measurements	Use Portable Multi-Gas Monitor	Describe Reciprocating Compressors	Prepare Maintenance Requests	
ROUTINE TASKS	Carry out Routine Equipment Checks	Change Process Filters	Describe and Change Corrosion Coupons	Monitor and Adjust Inhibitor Injection	Perform Housekeeping
SITE-SPECIFIC KNOWLEDGE AND TASKS	Describe Remore Process	Start and Adjust Remore Process	Describe and Change Remore Output Parameters	Perform Emergency Shutdown of Remore Process	Shut down Remore Process for Maintenance

Critical content for each competency is a list of the key issues a buddy or supervisor would emphasize when coaching the trainee. The end product is a *scope document* listing the key issues and ensuring continuity between competencies—no overlaps or gaps in content. As an example of a scope document, here is a partial list of key issues for the competency *Purge Piping and Station Systems*:

• replacing one medium with another to prevent combustible or toxic condition
• important to prevent:
  − people being exposed to toxic gases
  − possibility of a fire
• piping should only be purged after system has been opened and exposed to a foreign substance
• stations purged in preparation for startup
• some stations have automatic purging for specific piping and equipment
• automatic purging sequence must be checked
• always purge in direction gas migrates (up or down)
• criteria for length of time to purge include volume, pressure, and amount of connected equipment

In a profiling workshop, I used a brainstorming technique with four to sixteen of the customer's employees to identify competencies and critical content. The workshops were mentally demanding. On the one hand, I was concerned that the scope of training and performance requirements be limited and only address competencies and content that were considered important to the workers, their supervisors, and the organization. On the other hand, I was concerned that critical issues affecting people and the business were not overlooked. During these workshop sessions, I was constantly searching for relevant, useful, and practical content. What do the workers do? Is there a special way of doing the task? How do they know they are doing a good job? What can go wrong? How can the equipment be damaged or its life shortened? What do you mean by product quality? What about safety and the environment? Does the organization have special policies and ways of doing business? What is important and to whom or what? What questions should I be asking the group? I did not have a clear set of criteria or a structured thinking process that I could use to provide leadership in identifying training content that was important to the worker and the supervisor.

Working with Subject Matter Experts (SMEs)

I certainly believed that asking quality questions was more important than providing content. Answers to the questions could be provided by the customer's experienced employees. The term *subject matter experts* (*SME*s) is often used to refer to the organization's staff who provide content to training consultants and technical writers. Unfortunately, some SMEs, having in-depth knowledge of the tasks, technology, and the organization, had difficulties identifying content important for training. These SMEs expected consultants to provide leadership to identify relevant content. I soon discovered that my consultants often had difficulties in providing leadership to SMEs trying to identify content that was relevant, practical, and useful. When reviewing the first draft of training modules, information that would help trainees do their jobs more effectively, efficiently, and safely would often be missing. Nor would the supervisor's concerns always be addressed. Sometimes, information would be included that was of little value in helping workers do their jobs well and making decisions in the best interest of their organizations. When consultants asked me for direction as to the types of content that were relevant for training, I could not provide a comprehensive explanation. If the company was going to be successful in the future, I needed to find ways to define content that was relevant, practical, and useful—content that contributed to employee, job, and corporate performance.

Customer feedback gave me reason to believe that I was providing adequate leadership to identify relevant content; that I was asking quality questions. The truth of the matter was I did not have a formal list of types of question I should ask. In many ways, I was relying on intuition to ask the right questions. I needed to find a way to articulate a content gathering strategy that consultants could use with a variety of customers in different lines of business, different technologies, different hiring practices and performance expectations, and different ways of conducting business. I needed to find a way to identify the specific types of question consultants could ask SMEs to identify important training content—content that would help workers perform their jobs safely and effectively and contribute to meeting corporate objectives.

To help our training consultants and technical writers gain a better understanding of our customers, their businesses, goals, and concerns, I took consultants along to the competency-based profiling sessions. Listening to the group discussions and individual insights about the work and the business always provided learning beyond the information recorded in the program outline and scope document. This learning should be valuable when working with SMEs to identify detailed content for the training resources. Having this preliminary knowledge about the customer seemed to help some consultants be better at identifying relevant training content, but other consultants continued to struggle. I concluded that knowledge about the customer was valuable but didn't give consultants the strategies they needed to provide leadership when working with SMEs.

The Importance of Training Content Being Relevant to the Organization, Job, and Employees

Project reviews with customers were very useful for gaining ideas on how to improve services and products. Feedback from SMEs was that HDC consultants asked more questions than anyone they had ever worked with before. On the other hand, our consultants felt that they didn't ask enough questions because relevant information had been missed. The real issue was to ask fewer questions but more *quality* questions— questions that addressed issues that were important to employees, the job, and the organization. Certainly, customers strongly indicated that identifying relevant, useful, and practical content was the most important quality concern they had regarding the development of training resources. Customers also were adamant that consultants provide direction and leadership when working with SMEs to identify relevant content.

At the close of each project, I would ask the customer what additional training might be useful for consultants to help them be more effective at identifying relevant content. Suggestions included that consultants could increase their technical knowledge, or have a better understanding about safety management systems, environment management systems, or management styles. In response to suggestions, we began providing additional internal training using off-the-shelf technical training materials when possible. The additional training helped consultants to better understand what SMEs were telling them but only resulted in marginal improvements in consultants being able to provide leadership to identify relevant content. I concluded that the knowledge is useful but not sufficient in helping consultants (and workers) to identify issues important to employee, job, and corporate performance.

To compound the problem of identifying relevant content, expectations in industry were changing from developing entry level training (do as I tell and show you and don't ask why) to exemplary level training (maximizing productivity and making quality decisions) and every level between those two extremes. These changing expectations created difficulties in determining the content and amount of detail to include in training and keeping within training development budgets. Customers were upset if training materials included content they did not want and were not willing to pay for. Customers could also be disappointed if the training did not include content that they considered important. In many ways, the concerns consultants had in understanding the customer's expectations are the same concerns an employee new to a job would have.

When I had worked with exemplary workers, I discovered that one of their strategies was to confirm expectations. So we used the same strategy and built more confirmation checks into the development process to ensure the content was what customers wanted. Unfortunately, the confirmation checks were good at confirming that the documented content was what customers wanted but did not effectively address concerns about omissions of content important to customers (e.g., safety, equipment life).

Identifying Thinking Strategies Used by Exemplary Workers

Developing internal training for consultants to effectively identify relevant, useful, and practical content proved to be very difficult. Having consultants participate in the profiling sessions to learn about the customer, developing scope documents, providing technical and organization training, and building in confirmation checks had some value but weren't sufficient in helping them to provide leadership to identify relevant training content.

The instructional systems design models I was familiar with generally placed a strong emphasis on instructional development processes and only provided marginal direction and strategies on how to provide leadership to identify content that was important to customers. Certainly, the design of instruction and the nature of the content had an effect on each other. I suspected that there were instructional designs in which generic module structures and generic types of content would work for some types of technology and associated training outcomes. It would be several more years, after we had a large inventory of customized self-instructional modules, before we were able to develop a set of generic boilerplates (list of section and sub-section titles) for specific technologies and training outcomes. These *boilerplates* provided general structures for self-instruction and listed the types of content that *could* be included (but not necessarily included) in each section. No doubt, the SMEs that I worked with had mentally created their own boilerplates to be effective when working with specific types of equipment.

My initial effort to develop training to identify relevant content proved to be fairly impractical. Fortunately, several events provided me with the fundamental concepts needed to develop strategies that consultants could use to identify relevant content.

One of HDC's customers had a very demanding supervisor who was exceptionally analytical. In fact, he was by far the most powerful analytical thinker I have met. He was also driven to prevent anything negative from happening. He would always be analyzing situations and wanted to know all the *hows* and *whys* about every aspect of the instructional design that came to mind. Once a week I would make a personal visit to address his concerns. On one of those visits, he demanded to know what type of content should be addressed in the training. He said he asked our consultant the same question and the consultant's response was that *he would write self-instruction on anything as long as we told him the content.* Obviously, the consultant was not providing leadership when working with the SME to identify training content that would help the operators perform their work safely and effectively. For me, it was confirmation that our internal training was not very effective in helping our consultants to provide leadership.

My immediate response to his demand was to give some general criteria for identifying relevant content. *Well, safety, environment, equipment life, product quality, and customer satisfaction are important. Adhering to legislation and making decisions are important, too.*

There was a long silence—a lot of mental processing was going on in his head. Finally, he nodded and said, *Good. Let's tell the consultant and the senior operator what you just said.* The bottom line for this customer was that the training we were developing would contribute to his staff doing their work effectively and safely and making good decisions.

The interaction I had with that customer was the moment of discovery for me! The three-hour drive back to the office gave me time to reflect on what had just happened. Obviously, until I was asked, I had not been able to see the forest for the trees. Ask any business person what is important to their business success and he or she would give a list of areas of concern similar to the one I gave to my customer. No doubt the business person's list would be more extensive and include additional concerns affecting productivity and controlling losses—all businesses want to get the most out of their assets, including their people. Businesses prefer to have exemplary workers, workers that contribute to business success. Certainly, the training we develop for customers must help workers be effective in doing their jobs.

Creating the LO-PEMEO Model to Identify Relevant Training

I reflected on the thinking process I was using to identify relevant content when developing training profiles and scope documents. The questions that I had been asking myself during the sessions addressed the optimization and prevention of losses primarily to People, Equipment, Materials, Environment, and the Organization as a whole. Surely, the questions would take on meaning when the work environment was considered. And one way of assessing the work environment was to consider the conditions, actions, and events within the workplace that affect PEMEO.

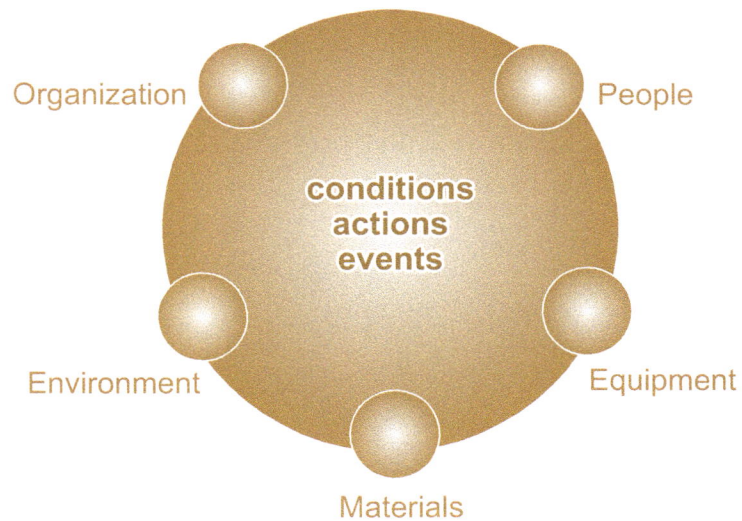

Most exciting for me, I could combine the concepts of optimization and controlling losses of organizational assets such as people and equipment to create a model and strategy for identifying relevant content. The LO-PEMEO model was born. Each of the five domains (people, equipment, materials, etc.) shown in the above illustration had potential for optimization and loss. An example of loss to people is illness and injury. Loss of materials when processing ore is the inefficient recovery of the desired products. Optimization of materials in construction is to use the right materials and maximize the use of the materials. The following illustration shows the combinations of loss and optimization of PEMEO.

	LOSS				OPTIMIZATION	
Loss:	People	LP	P	OP	Optimization:	People
Loss:	Equipment	LE	E	OE	Optimization:	Equipment
Loss:	Materials	LM	M	OM	Optimization:	Materials
Loss:	Environment	LE	E	OE	Optimization:	Environment
Loss:	Organization	LO	O	OO	Optimization:	Organization

Exemplary workers consider the potential for Loss and Optimization of each domain of PEMEO (i.e., LO-PEMEO) while they work. So LO-PEMEO was used as the framework and structure for *The Exemplary Worker* series of books. For example, loss to people (LP) is safety—the book *SafeThink* focuses on using a structured critical thinking strategy to identify and predict hazardous situations to prevent illness and injury.

Interestingly, several years later, I was introduced to a loss control model created by Frank E. Bird that used PEME as an acronym. I have always wondered if it would have saved me a lot of effort if I had known of Bird's loss control model earlier. Or would that knowledge have put in place constraints such that I would never have created the LO-PEMEO model?

While driving back to my office, I thought about how fortunate I had been over the years to work with a lot of exemplary performers, many of them my SMEs. Our customers gave us SMEs who are exemplary workers because the belief is that exemplary workers know what is important for business success and will provide training content that is relevant to corporate, job, and employee performance. When I had asked the SMEs if there were any concerns about issues such as safety, equipment, or materials, they would often look at the ceiling and ponder for a while. If they said yes, they would go on and give me further clarification. If they said no, I would continue to ask different questions. When I thought about it, the questions that I asked SMEs usually focused on concerns about LO-PEMEO. I always wondered what the SMEs were thinking when they were looking at the ceiling and pondering the answers to my questions. Eventually, I asked them. Interestingly, different SMEs from different companies and lines of business had similar concerns. For example, damage to equipment often involved shock from a sudden change in physical forces or temperature. The sources for causing damage could be people, material, or any of the other three domains. In fact, *each domain has the potential to affect the other domains*. Whether the SMEs were aware of it or not, they were mentally searching for specific workplace concerns relating to LO-PEMEO. In many ways, even at the detailed level, *the thinking strategies of exemplary workers were similar and generic*. Certainly, being aware of one's own thinking strategies contributes to planning and working effectively and helps to communicate effectively when collaborating with others and mentoring apprentices.

Linking Corporate, Job, and Employee Performance

When organizations develop standards, procedures, and training, they want to realize an improvement in corporate performance. Improving *corporate performance* is often achieved by either filling a gap in performance or by preparing the organization to move towards new goals. The following illustration lists some criteria that can be used to measure corporate performance.

PERFORMANCE REPORT

Customer Satisfaction	UP
Production	UP
Product Quality	UP
Equipment Run Time	UP
Equipment Damage	DOWN
Energy Consumption	DOWN
Material Waste	DOWN
Personal Injuries	DOWN
Maintenance Costs	DOWN
Environment Damage	DOWN
Rework Time	DOWN

At the operational or job level, the supervisor also has concerns about performance. Within his or her roles, responsibilities, and authority, the supervisor is expected to maximize productivity and minimize losses. Improved *job performance* contributes to improved corporate performance. The supervisor therefore represents the concerns and goals of the organization and must use specific resources and assets (including people) to effectively achieve the goals. The supervisor must also be able to motivate, coordinate, and assign staff to effectively carry out the work. Furthermore, worker performance affects job performance which, in turn, affects corporate performance.

Employee performance affects business results. Employees are expected to work effectively and efficiently and make good use of materials and technology. Expectations of performance are articulated to line employees both orally and in writing. In turn, employees have concerns about understanding the expectations and working safely, effectively, and efficiently to meet the expectations. The following illustration is of a person new to a job asking questions relating to corporate, job, and employee performance issues.

> What's important to the business?
>
> What does the team leader expect of me?
>
> What am I supposed to do?
>
> How am I supposed to do it?
>
> How do I know I've done well?
>
> How does my work affect others?
>
> Is there a better way?
>
> What tools and equipment are used?

> Could I get hurt?
>
> Could I injure others?
>
> Could I damage the equipment?
>
> Does this product affect the environment?
>
> How much waste is acceptable?
>
> How can I prevent…?
>
> Will the customer be satisfied?
>
> What should I do if …?

> What would happen if …?
>
> Do I have the authority to take action?
>
> What action?
>
> Whom should I inform?
>
> What does …?
>
> How does …?
>
> What caused …?
>
> What is the reason?
>
> What are the consequences for …?

> What questions should I be asking?
>
> What answers do I need?

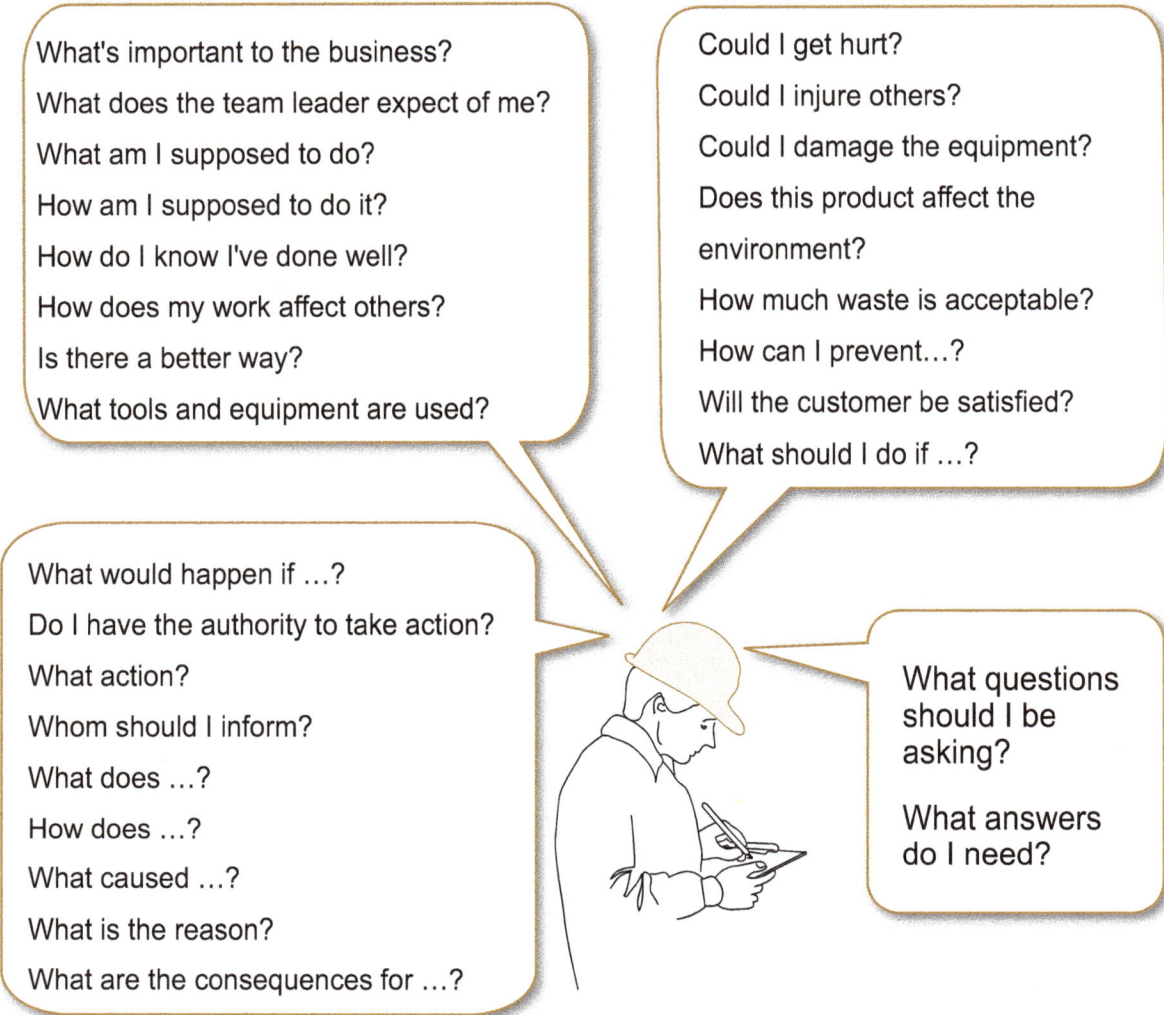

Many of the questions are generated by the LO-PEMEO strategy and focus on performance:

- What is important?
- What are the issues?
- What questions should I ask?

The person new to the job needs answers to the questions in the illustration to quickly learn to do that job effectively and efficiently. Interestingly, two employees with similar experiences and skills who are new to a job can perform quite differently. One employee will be uncertain about the work and become stressed if work conditions change. The other employee will initiate actions and make good work-related decisions for the organization within a few weeks. One of the factors that makes the difference in performance between the two employees is the knowledge about what is important to job and corporate performance. Understanding *what is important* provides criteria for focusing one's efforts and for making decisions. LO-PEMEO is a good start in identifying what is important to the organization. Although many of the issues identified by LO-PEMEO are generic, each organization has its own business strategies, resources, and priorities. As such, each organization could place a different emphasis on each issue identified by LO-PEMEO. And that's why asking the *right* questions is so valuable. Questions focus on key issues; the answers to the questions are unique to the organization, workplace, and specific circumstances. *The Exemplary Worker* series provides many of the questions that workers need to ask of themselves and of others to achieve exemplary performance.

Understanding Organizations for Exemplary Worker Performance

Exemplary workers understand what is important to the organization so that they put their efforts in the right places, do the right things, and make good decisions in the best interests of their organizations. For workers to have exemplary performance, they need to have an understanding of organizations in general, and a specific understanding of their own organization. Training and performance consultants also need to have a general understanding of organizations to be effective at developing customized training—training that is relevant, useful, practical, and reflects the organization for what it is. There is a lot of literature on organizations but most of it is more complex than training consultants need. Generally, the literature does not directly address issues important to designing and developing customized training for industry.

So, what issues are important? For consultants at HDC (and exemplary workers in other organizations) to be effective, they must be able to identify and understand organizational issues from different points of view. Imagine a roomful of statues facing in different directions. The room has many doors, each opened by a different work group or discipline. Each doorway has a different view of the statues.

For consultants to get a broader understanding of the organization, they need to view the statues from different doors. Ideally, consultants would walk around the statues to get many different points of view. The consultant must be prepared to consider different points of view within a specific organization to be effective at understanding the organization and identifying issues important to employee, job, and corporate performance.

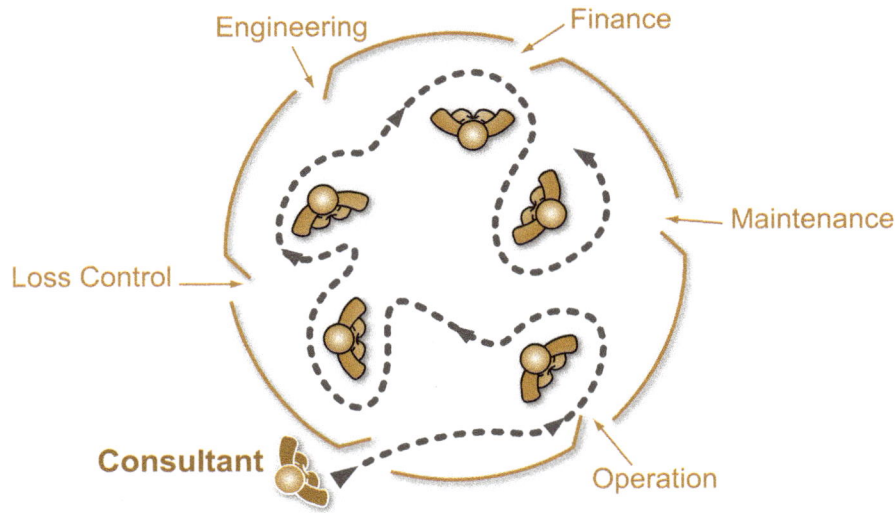

Both exemplary workers and training/performance consultants benefit from an understanding of relationships between business resources, organizational structure, business strategies, corporate objectives, and performance standards. Exemplary workers gain an understanding as to how their line of work fits into the organization as a whole. In doing so, they appreciate how their work affects others and they potentially make better use of organizational resources. This understanding about organizations also helps training consultants and technical writers to be more effective at designing and developing training that is customized, reflects the business, and has excellent value for the customer.

The approach I take with consultants to learn about organizations is to pretend to build a new business. Would the line of business be a service or a product? What is the mission? If the business is a service, then performing tasks is the main way to generate revenue and tools/equipment provide support for carrying out the work. If the line of business is to use technology to make products, then the technology dictates many of the tasks that workers must do. Having resources to achieve specific results is essential but not sufficient for business success. The resources must also be managed effectively. The following illustration identifies some key constituents of a business.

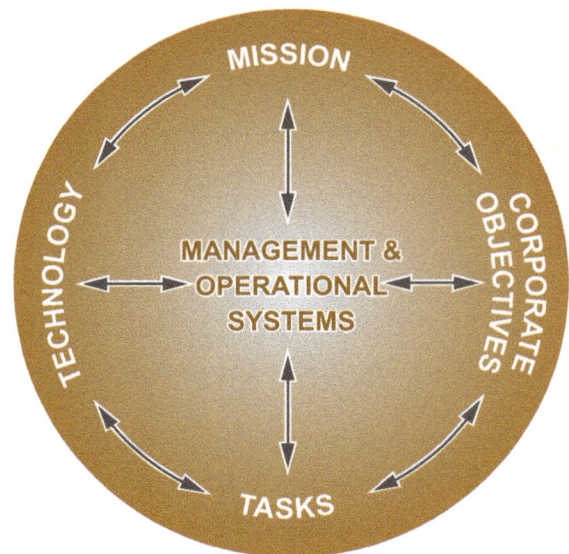

The book *JobThink* uses the previous model to provide a practical way for workers to understand organizations. This understanding helps workers to effectively focus their efforts and make decisions in the best interests of their organizations.

Of particular interest are the *corporate objectives*. Corporate objectives provide direction for using technology, performing tasks, and coordinating work to effectively achieve the corporate mission. The following table lists areas of concern, common to many organizations, for which corporate objectives may be developed.

Areas addressed by Corporate Objectives

- safety
- environment
- legislation
- equipment reliability and life
- equipment optimization
- energy use
- quality
- waste control
- loss control
- cost control
- customer satisfaction
- public image
- public disruption
- reputation
- communication
- teamwork

For a specific organization, a list of corporate objectives can be generated by expanding the organization's strategic business objectives or by using LO-PEMEO. Some companies issue strategic business objectives to provide direction to employees as to where to put their energy and focus for business success. Strategic business objectives identify what the organization must do well to be successful. For example, leaders in an organization may believe that it is essential for business success to have reliable service and satisfied customers. Organizations may identify five to eight strategic objectives. Within a department, the list of objectives (or goals) may be expanded in more detail to address issues specific to the department's mandate.

The expanded list of corporate objectives can also be generated using LO-PEMEO—each of the items in the above table relates to one or more of the LO-PEMEO domains.

Corporate objectives are fundamental to exemplary performance because they define what is *important* to the organization, the job, and workers. Corporate objectives provide a ***formal link*** between organizational goals and worker performance. Workers can use corporate objectives as criteria for working effectively and efficiently and for making decisions in the best interest of their organizations. Training consultants and technical writers can use corporate objectives to identify relevant, useful, and practical training content. Refer to my book, *Interviewing to Gather Relevant Content for Training* for:

- information about applying critical thinking skills to identify relevant content for training
- an interviewing process that consultants and technical writers can use to interview SMEs to gather relevant content

Developing Training to Identify What is Important to Employee, Job, and Corporate Performance

With the LO-PEMEO and business models, I could now develop training for consultants to provide leadership to identify relevant content. The LO-PEMEO model was the most practical approach to use to structure the training because it relates directly to work and job issues. The organizational model can be integrated into the training on loss and optimization of organization, LO-O. For the training on these models to be useful, the training needs to be flexible and apply to a broad range of work, technology, and organizations. The training must also provide strategies for people to think through their work. That is what exemplary workers do—they think through their work. And, the thinking processes are generic so they apply to all types of industries, work environments, and jobs.

All of the training to identify relevant content is founded on using thinking strategies. An emphasis is placed on *concepts* and *generalities* to maintain a broad application of the thinking strategies. Furthermore, the thinking involves asking questions relating to LO-PEMEO. Asking questions is important to maintaining the broad application of the thinking strategies and helping people remain mentally engaged. Asking the *right* questions is often more important than finding the answers, because if the right questions are asked, answers can usually be found—answers that contribute to exemplary employee, job, and corporate performance.

Over several years, I developed training for all the combinations of LO-PEMEO. I also expanded the training to include consulting processes and a performance and training model to design, develop, and implement competency-based training and performance management systems. I was very fortunate to have excellent support from staff to edit and refine the training. HDC staff made important contributions to the training content and presentation. And, after the training resources were in use, we refined them further.

Developing *The Exemplary Worker* Series

After the HDC consultants' training resources had been used for ten years, I decided to go full circle and modify the resources for general use. A major rewrite was required; the new audience was very broad and the lines of work very diverse. The instructional design content had to be deleted. New and different examples of applying the thinking strategies were required for the books. To help the reader, each book required new learning activities. Exemplary workers in industry needed to field test and validate the content. Staff also needed to make major contributions to ensure the quality of each book. It took over six thousand hours to develop *The Exemplary Worker* series. In addition, industry has volunteered more than a thousand hours to field test and validate the content.

The Exemplary Worker series has many suggestions to help you not only be aware of your own thinking strategies but also help you to refine your strategies to achieve exemplary performance. You will also be better at mentoring others to perform better.

Gordon D. Shand
Edmonton, Alberta
Canada

Training Objectives

Upon completion of this book, you will be able to apply a thinking strategy to identify what is important to using tools and equipment effectively so that you achieve the desired results. You will be able to:

- Identify critical **input variables** of materials that define the quality of the materials used
- Identify the critical work and technical **process variables** that affect the quality of results
- Identify critical **output variables** that define the quality of results
- Optimize the use of tools
- Select mobile equipment for the job
- Use output, input, and process variables to operate mobile equipment effectively
- Identify critical output, input, and process variables associated with stationary equipment
- Identify critical variables associated with stationary equipment operation and condition
- Operate stationary equipment effectively
- Make changes to work and technical processes in response to:
  - desired changes in the quality of results
  - changes in input materials
  - abnormal conditions
- Describe three ways to assess equipment life
- Describe maintenance strategies to maintain equipment reliability and extend equipment life

Introduction

This book is one of *The Exemplary Worker* series of books. Books in the series all focus on using critical thinking strategies to identify ***what is important*** to employees, the job, and the organization. Each book focuses on one of five domains (**PEMEO**):

P People

E Equipment

M Materials

E Environment

O Organization

Within each book, loss and/or optimization (LO) are the main themes, hence the word LO-PEMEO™:

Themes	Books
L-P Loss to People (Safety)	*SafeThink* Use a structured thinking strategy to identify and predict hazardous situations.
O-P Optimize People's Performance	*WorkThink* Work effectively and efficiently.
LO-E Loss and Optimization of Equipment	*EquipThink* Use tools and equipment effectively and efficiently.
LO-M Loss and Optimization of Materials	*MatThink* Use materials effectively and efficiently.
LO-E Loss and Optimization of the Environment	*EnviroThink* Protect the environment.
LO-O Loss and Optimization of the Organization	*JobThink* Contribute to job and corporate performance.
LO-PEMEO Use thinking strategies for the workplace	*MetaThink* Integrate thinking strategies for exemplary performance.

The fundamental premise of LO-PEMEO is to *ask questions.* By asking yourself questions, you remain alert. By seeking answers, you continually learn and become more effective in the workplace and adaptable to changes. The big question is: *What questions should I ask?* The questions identified in LO-PEMEO help you to ask many of the right questions.

This book *EquipThink* focuses on loss and optimization (LO) of tools and equipment. Damage to equipment and shortening equipment life are two examples of loss to equipment. Operating equipment efficiently, obtaining the desired results, and minimizing stress on equipment are three examples of optimizing equipment. Generally, if you optimize equipment operation, you reduce losses to equipment and vice versa. The Job Aid for this book lists the key questions you can ask yourself to select and use tools and equipment effectively. The lists are quite extensive. For each list, check off the questions that apply to your work and equipment. Add any questions that may be work- or equipment-specific.

The learning activities that are integrated into this book provide you with opportunities to apply critical thinking strategies to your job, workplace, and personal activities. Each learning activity relates to specific concepts addressed in this book. It is recommended that you complete each learning activity as you progress through the book.

In this book, the word *tools* refers to hand and portable power tools. The word *equipment* refers to both mobile and stationary equipment. In all industries, tools and equipment are an important part of doing business. It is important for business success to:
- use tools and equipment effectively to achieve the desired results
- ensure the reliability and life of tools and equipment
- conserve energy

Most, but not all, tools and equipment create change. For example, some tools and equipment:
- change material physical characteristics such as shape and finish
- apply or remove materials
- change material composition

Some tools and equipment (e.g., magnifying glasses, electrical generators) do not create change to materials. This book primarily focuses on tools and equipment that change materials. However, some concepts addressed in this book also apply to tools and equipment that do not change materials.

Tools, equipment, and materials are closely related. The function of tools and equipment is to create change to materials. Conversely, materials can adversely affect the condition and life of tools and equipment. Controlling the condition and life of tools and equipment can be difficult because several domains of PEMEO other than materials can cause harm. For example:

- people (**P**) can often use tools and equipment for applications that they have not been designed for (e.g., using a screwdriver as a chisel). People can also operate the equipment in ways that can cause harm to the equipment condition and shorten equipment life.
- the environment (**E**) can cause components to deteriorate (e.g., rust)
- the organization (**O**) may use inadequate maintenance practices

Tools and equipment can also affect PEMEO domains other than materials. For example:

- people (**P**) who use the tools and equipment are at risk of injury
- some equipment (e.g., internal combustion engines) create emissions that harm the environment (**E**)
- poorly functioning equipment reduces productivity (**O**)

Where applicable, this book addresses both how PEMEO impacts tools and equipment and how tools and equipment impact PEMEO.

The two most critical factors affecting the optimum use of tools and equipment are:

- selecting the right tools and equipment for the job
- using tools and equipment effectively and efficiently

Change Process

When working with tools and equipment, your main concern is to control changes through work and technical processes to achieve the desired results (quality). The following diagram shows the three stages of the change process:

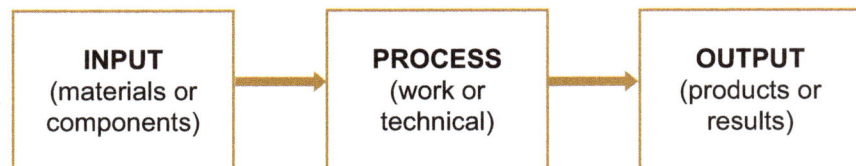

INPUT (materials or components)	→	**PROCESS** (work or technical)	→	**OUTPUT** (products or results)

As a generality, work and technical processes are used to change materials to produce a desired result (e.g., reshape metal, cut glass to size, assemble an appliance, separate hydrocarbons into various components, create a chemical change). Generally, for a given project or customer, it is desirable for the quality of the output product to remain constant. Sometimes a change in the quality of the input materials creates the need to adjust the work or technical process to achieve the desired results.

In this book, the terms *work process* and *technical process* have different meanings:

Work process: people carry out a task to achieve a specific result. Often tools and equipment are used as part of carrying out the work. The task, the types of material, and the desired results determine the tools and equipment that will be used to carry out the work. For example: many types of lawn mower can be used to cut grass; many different tools and equipment can be used to cut metal; many different types of earthmoving equipment can be used to move dirt.

Technical process: a stationary piece of equipment for manufacturing or processing performs a specific function to make a change to materials. The specific equipment dictates the tasks that people must do to operate the equipment. For example: start, monitor, adjust, shut down, and maintain the equipment.

1.1 Variables in the Change Process

Material properties: *scientifically defined qualities such as composition, density, and melting point*

Material characteristics: *size, shape, surface finish, type of break that can often be controlled by the manufacturer*

Each stage of the change process involves variables:
- **input variables** define the quality of the materials used
- **process variables** of tools and equipment that create the change
- **output variables** define the quality of the product or results

Measuring Variables

Variables relate to properties, characteristics, or settings. Units of measure are used to define the quality of the input, process, and output variables (e.g., temperature, pressure, volume, mass).

Variables can be measured numerically or non-numerically:
- **numerical measurements** may be expressed as a unit value or ratio:
  - units of measurement (e.g., 2500 revolutions per minute, pressure of 100 kPa, power of 3730 watts or 5 horse power)
  - ratio or percentage (e.g., 1:15 ratio of fuel to air, 7% asphalt by weight)
- **non-numerical measurements** are expressed in terms of our sense:
  - a liquid appears cloudy instead of clear
  - the surface of a board appears rough instead of smooth
  - apply enough force to obtain a clean cut but not enough to overload the motor

Input, process, and output variables have target values or optimal ranges. For example:
- the lumber must be less than 50 mm thick
- use a #2 Robertson screwdriver
- set the rpm to 1200
- torque the bolts to 30 foot pounds
- control the pressure of the gas entering the vessel between 500 kPa and 575 kPa
- the width of the material must be 20 mm, + 1mm
- the optimal composition of the product is 40% A, 60% B

Classification of Variables

A variable can be classified according to its state and controllability:
- either static or dynamic (state)
- either controllable or non-controllable (controllability)

static variable	a variable that cannot be easily changed. For example, the hardness of oak lumber.
dynamic variable	a variable that can easily change. For example, the velocity of the wind, the number of orders of a product.
controllable variable	a variable that you or technology can change. For example, the rpm of an engine, the depth of cut of a milling machine, the dimensions of plywood parts.

non-controllable variable   a variable that cannot be easily regulated. For example, the frequency of customer orders.

The two categories (state and controllability) of a variable combine to produce four possible types of variable as shown in the following illustration:

Variable	Static	Dynamic
Controllable	static and controllable	dynamic and controllable
Non-Controllable	static and non-controllable	dynamic and non-controllable

Knowing whether an equipment or material variable can change and whether or not you can control the change is very helpful in doing your job well and responding effectively to changes. You must also understand the cause(s) or reason(s) for changes and how changes can impact PEMEO (i.e., LO-PEMEO).

The *type* of variable can have a significant impact on the loss and optimization of PEMEO and on the decisions and responses that you make (within your limits of authority). The type of question you ask varies with the type of variable. For each of the following examples, the questions in italics are the types of question you might ask yourself to work effectively and efficiently and make *good* decisions.

Static and controllable variable: This type of variable remains the same throughout the work or process cycle but can be changed with some effort. Usually production must stop to make the adjustments. For example, the power of a gasoline engine is fixed, however, mechanics can modify the engine to increase or decrease the power output. The type of band saw blade (number and size of teeth) used to cut logs into lumber can be changed. The operator must use the original blade during the work cycle. During use, the blade becomes dull, resulting in inefficient cutting and an increase in blade temperature. During the shutdown cycle, the blade may be changed to one that is sharp or functions better for the type of wood being sawed. Ask yourself questions about how the change affects the operation and the materials. Answers to the questions help you adjust the way you operate to achieve production goals and minimize losses.

Ask yourself: *How dull must the blade be before it is changed? How do I know the blade is dull enough to require changing? What type of blade is used for each type of wood? Are blade changes made to match with the type of wood being processed? Who makes the decision? If the type of wood entering the saw mill often changes, what work adjustments must be made to maintain production, control quality, minimize stress on the equipment, and prevent personal injury? Does maintenance make adjustments to the saw while it is running?*

NOTE

A variable may be static or dynamic, depending on the job position or point of view. For example, at a specific maximum pressure, a piece of equipment will shut down. For the operator, that variable is static and does not change. An instrumentation technician, however, views the variable as dynamic because he or she can change the shutdown pressure and *fix* it at a new setting. For the operator, the new setting is *static* and he or she must work with the new setting until the technician makes a further change.

Dynamic and controllable variable: This type of variable can change throughout the work or process cycle, however, there is a means of controlling the variable to prevent damage to equipment and materials and to achieve the desired results. For example, the pressure of raw gas feeding to a specific vessel fluctuates. Because the vessel requires a specific gas pressure to function properly, a pressure regulator is installed on the line. The regulator continuously makes adjustments to ensure the gas pressure in the vessel remains within a specific pressure range. The regulator may be designed to be manually adjusted to obtain different regulated gas pressures. Asking yourself questions and seeking answers about the dynamic and controllable variables helps you work more safely and effectively.

Ask yourself: *What is the acceptable pressure range of the gas entering the vessel? What is the impact on PEMEO if the pressure goes outside of the desired range? What is the desired response if the pressure is unacceptable? Who decides to take action? If the pressure goes outside the operating range, is the product leaving the vessel still within acceptable specifications? What should be done with off-spec. product? Who decides? Who takes actions? Does the incident have to be documented?*

Static and non-controllable variable: This type of variable does not change and cannot be changed. The composition of ore from a specific part of a mine remains constant. Ore processing equipment may have to be adjusted to effectively process that specific ore composition.

Ask yourself: *What impact does the ore entering the process have on PEMEO? What equipment processing variables can be changed to obtain efficient and effective processing? What are the specific settings? Who makes the decision to change equipment operation? Who makes the changes? How are the changes made? After the changes are made, how long will it be before there is a noticeable difference in the specifications of the output products?*

Dynamic and non-controllable variable: This type of variable changes and cannot be controlled, for example, the variables associated with weather. Variables associated with weather include temperature, atmospheric pressure, humidity, precipitation, and wind velocity. In industry, the composition, characteristics, and properties of raw materials may be dynamic and non-controllable (e.g., moisture content of incoming vegetables in a food processing facility).

Ask yourself: *What is the impact of the variable on PEMEO? Can work, processes, and equipment be adjusted to safely and efficiently obtain the desired output results? What adjustments can be made? Who has authority to make decisions? What types of decision are made? Who carries out the adjustments? How are the adjustments made? If adjustments are made, what are the impacts on PEMEO? Is there a need for communication before, during, and after the adjustments are made? Is any documentation required? Who does the documenting and how is it done?*

NOTE Often the type, size, and quality of large, stationary equipment for processing is static but some material, equipment, and processing variables may be controllable to achieve the desired output results.

The questions previously listed in italics are the types of question exemplary employees might ask when confronted with a specific work condition. Identifying variables, determining their types (change and control) and determining the impact that variables have on PEMEO are important thinking questions that you need to ask and answer to work effectively and efficiently. Within your roles and responsibilities and limits of authority, the answers to these types of question also help you make *good* decisions and respond effectively to both static and dynamic variables.

Variables can be a powerful way to identify important issues for many different tasks and technologies. For example, using word processing software to format paragraphs involves several variables, including:
- space before the paragraph
- line length (gutter, left margin, and right margin)
- font type
- font size
- space between lines
- space after the paragraph

1.2 The Thinking Strategy

This book provides a **thinking strategy** to answer the question, *What is important about tools and equipment in my job?* The emphasis is not on the features of tools and equipment nor on the properties and characteristics of materials. Instead, the book focuses on the thinking strategy of *input, process,* and *output* variables so that you can select the right tools and equipment for the job and use them effectively to achieve the desired results. The thinking strategy also helps you to respond more effectively to changes in equipment status and condition.

Using *variables* associated with input, process, and output can help you:

- better understand how work and technical processes create change
- determine the functions, applications, and limitations of tools and equipment
- identify the input variables of materials that must be changed by the tools and equipment
- identify the input variables of materials that must not be changed by the work or technical process
- identify the input variables of materials that can affect the condition of tools and equipment
- identify the variables of tools and equipment that create the change
- determine the variables of tools and equipment that you control to achieve the desired results
- determine which variables of tools and equipment you cannot change but which can affect the quality of results
- identify operating variables that affect equipment condition and life
- define the quality of output results
- adjust manufacturing and industrial processes more effectively in response to input variable changes to maintain the desired quality of results
- adjust input variables and/or processes in response to a desired change in the quality of results
- use tools and equipment effectively to achieve the desired results and minimize energy consumption
- identify variables associated with tools and equipment that can cause harm to the environment
- identify how tools and equipment can affect PEMEO
- effectively communicate critical work and technical process issues to others

IMPORTANT

The *input, process*, and *output* thinking strategy helps you focus your thinking to identify variables that are important to your job. After you identify the important variables, you need to ask questions and seek answers about the variables so you can do your work safely, effectively, and efficiently.

This book applies the concept of input, process, and output variables to:
- selecting hand and portable power tools
- selecting and operating mobile and stationary equipment

Optimize the Use of Tools and Equipment

In this section, a strong emphasis is placed on using the *input-process-output* thinking strategy. The following sections focus on the applications of the thinking strategy.

This section focuses on **selecting** the best hand or power tool to make a change to materials or to attach/detach components for operation and maintenance. The proper use of tools (i.e., the skills) is beyond the scope of this book.

In your job, using the *input-process-output* thinking strategy is very useful to determine the specific variables affecting the selection and operation of tools and the quality of results.

For work processes:
- **Inputs** are the materials being used
- **Process** is using tools and equipment to create change
- **Output** is the result of the work process

There are many variables to consider. Variables for *input* materials and components include: strength, hardness, dimensions, shape, finish, and composition. Variables for selecting the best tool to carry out the task (work *process*) include: function, application, and size or capacity. Variables that create the change (e.g., cutting action of a saw blade) and variables that affect tool condition must also be considered. The variables defining the quality of results (*output*) must be clearly understood. However, it is neither practical nor useful to identify *all* of the variables for selecting (and using) the best tool for the material or component, and the task. You only need to identify the quality variables that relate to your work and affect your organization's goals (e.g., safety, quality, and quantity of the work performed).

The input, process, and output thinking strategy is used in three different sequences:
- input–process–output
- output–process–input
- output–input–process

You can think of input, process, and output as a continuous loop.

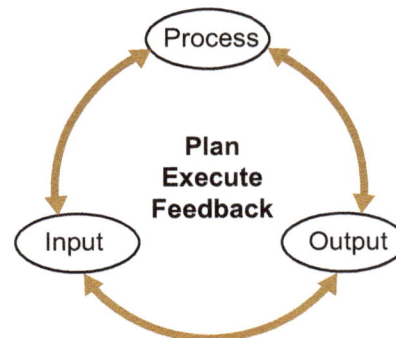

You can start at any point in the loop and go in either direction to plan, execute, and get feedback. After you have used the strategy to initially think through the work, you may find yourself then using a different starting point and sequence to get a better understanding of a specific concern. The goal is to continually refine and improve safety, efficiency, and cost-effectiveness in achieving the desired results.

In this book, when referring to the input-process-output thinking strategy as a generic strategy, it is written this way:

input–process–output thinking strategy

When applying the input-process-output thinking strategy,

the specific starting point and sequence is written in italics. For example:

output–input–process thinking strategy

Input–Process–Output

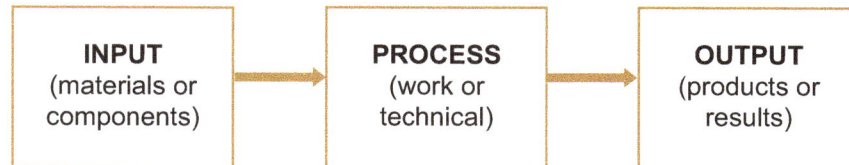

INPUT (materials or components)	→	PROCESS (work or technical)	→	OUTPUT (products or results)

In the illustration above, the key *input* variables that you must consider when selecting the best tool are material or component variables that:

- must be changed (size of a panel)
- must **not** be changed (surface finish of the panel)
- affect the efficiency of the work process and tool condition and life (fine-toothed saw blade on a portable power saw will become excessively hot if used to cut solid lumber)

The key work *process* variables are:

- tool-related variables that create the change
- material or component variables that are changed
- variables that affect tool condition

The key *output* variables are material or component variables that define the quality of results.

NOTE

The output of one work process may become the input material or component for the next work process.

Output–Process–Input

Sometimes, it is helpful to determine the critical variables by working backwards when there is limited choice in the selection of the tool to do the work (e.g., a portable circular saw must be used to cut material and there are a set of blades that can be used).

OUTPUT (products or results)	→	PROCESS (work or technical)	→	INPUT (materials or components)

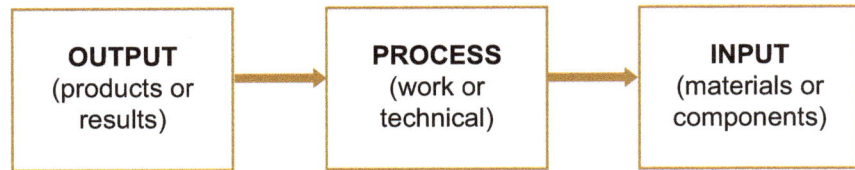

Using the portable circular saw as an example, first determine the material output quality variables. Next, identify the application of each saw blade (for crosscutting, ripping, cutting laminate, etc.) and how well the saw blade performs when doing the cutting. You need to consider the changes made with the cut (e.g., wide or narrow cut, clean or rough cut). You must also consider how a saw blade can cause damage to the material (e.g., by chipping the surface along the cut). Finally, examine the input material to determine which input variables must change (i.e., dimensions) and which input variables must not change (e.g., smoothness of cut, melting of the material). The information you have gained in the thinking process helps you select the most suitable saw blade available for the work. The information also helps you think about how you are going to carry out the task to get the best results (e.g., when cutting some plastics, a travel speed that is too slow can cause melting and a travel speed that is too fast can cause chipping of the edge).

Here is an example of using the *output–process–input* sequence:

output: Steel plates have four holes drilled near each corner to receive bolts. The location and size of the holes are critical for the bolts to align and hold firmly. Hole location and size depends on the specific size of plate being manufactured.

process: A drill press is used to make the holes. A jig attached to the drill press is used to align the steel plates. There is a different size drill bit and jig for each size of plate. Over time, both the drill bit and the jig can wear. For each size of drill bit, there is an optimal rotational speed (rpm) and downward pressure to operate effectively.

Material variables that are affected are:
- cleanness of the cut
- material characteristics must remain the same (excessive heat can cause the material to weaken)
- location of the holes

Because the holes are not 45° to the corner, the plate must be flipped over to drill two of the holes.

Equipment variables that must be controlled include:
- drill bit rpm
- downward pressure on the drill bit
- cooling of the drill bit and material while drilling
- selection and quality of the jig
- orientation of the metal plate to the drill press

The quality of the bits used for drilling affects the:
- price per bit
- quality of the drilled holes
- rate at which the holes can be drilled
- number of holes the bit can drill before it needs sharpening
- total number of holes the bit can drill before it must be replaced

input: The size of plate affects the location and size of the holes. Material properties affect the selection of the drill bit and operation of the drill press. The steel may be hard or soft. Different bits are required for different hardness of steel. The rpm and pressure on the bit may have to be adjusted for different material properties.

Both the materials and the equipment can affect other domains of PEMEO. The rotating drill bit can snag loose clothing and hair. Repetitive motion could cause soft tissue injuries. The cutting oil and the heavy metals in the cutting oil may have a toxic effect on the:
- operator (if absorbed through the skin or the vapors inhaled)
- environment when the used cutting oil is disposed of

Output–Input–Process

Industries such as construction plan carefully to ensure the right tools, equipment, and materials are selected for the job. For planning, often output quality variables are examined first and then the input quality variables are examined. The most effective tools and equipment for the job can then be selected. This thinking sequence can also be helpful in understanding how a technical process works to achieve the desired results.

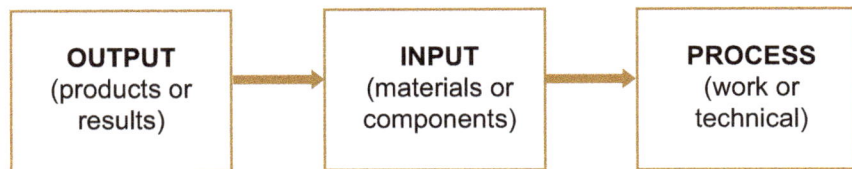

OUTPUT (products or results)	→	INPUT (materials or components)	→	PROCESS (work or technical)

Here is an example of using the *output–input–process* thinking strategy to plan and build a feature wall in a house:

output: The criteria (e.g., aesthetics and function) for designing the feature wall must be determined. Next, the type of material that will be used is selected. A layout (drawing) of the wall is made.

input: The properties, characteristics, and dimensions of the material are examined to ensure that they are suitable for the application. In some cases, the design must be modified to be functional and cost-effective. For example, material strength of the shelving material may require that the span of the shelving be reduced or different material used.

process: Given the desired results and the materials to be used, tools and equipment are selected that can do the job effectively and efficiently. Sometimes, the design or the materials may have to be changed because it may be very difficult to make the desired product using the available tools and equipment.

The *output–input–process* thinking strategy is then used to make the components of the feature wall. For each component, the desired quality of results is first determined. Next, the critical variables of the input materials are determined to select tools and equipment that will effectively achieve the desired changes to the materials. The critical material variables such as finish that must **not** change in the *process* are identified. Work processes are then established to do the work effectively and efficiently.

Using the *output–input–process* thinking strategy can be helpful in selecting the best tools to do the work effectively and efficiently.

Here is an example of using the *output–input–process* thinking strategy to select the best tool to cut a 28 mm (1.5 in.) diameter steel bar to length.

output:
   Criteria defining the quality of results are the:
- precision to which the length of the bar must be cut
- quality of the cut (e.g., rough, smooth, square to the axis of the bar)
- importance of maintaining the properties and characteristics of the steel
- time needed to make the cut

input:
   The hardness of the steel must be determined. Some tools may not be able to cut the steel; other tools may create enough heat to change the hardness of the steel. Some tools may cause scarring of the bar's surface.

process:
   Given the desired quality of results, the materials to be used, and the properties and characteristics of the input materials, several types of steel cutting tools can be considered. For example:
- **power hacksaw:** The cut will be fairly accurate and square. The cutting process is slow. Some grades of steels are too hard for the hacksaw to cut.
- **cutoff disc:** A stationary cutoff disc can make the cut quite accurately and quickly. Using a hand-held cutoff disc can result in a less accurate

cut. Cutting too fast could overheat the steel and change its properties. The clamping jaws could potentially scar the steel.

- **acetylene cutting torch:** The cut is quite rough but can be done quickly. The heat from the torch could change the properties of the steel.
- **steel lathe:** The cut can be made very accurately and square. Depending on the hardness of the steel, cutting time would be slower than when using a cutoff disc but faster than when using a power hacksaw. Overheating could occur. Cutting oil can be used to prevent the steel from overheating and facilitate the cutting.

Sometimes, two tools may have to be used to achieve the desired results: For example:

- to maximize production, an acetylene torch could be used to cut the stock steel close to the desired length. A grinder can then be used to improve the finish and accuracy of the cut.
- a steel lathe could be used to cut the steel bar to length but the stock steel is too long. An acetylene torch could first be used to cut the stock steel to rough lengths to fit in the lathe.

It is often advantageous to apply different sequences of input, process, and output to refine your understanding of the variables important to your job. By doing so, you improve your ability to focus on critical variables so that you are more effective at optimizing the use of tools and equipment.

NOTE

The use of *variables* and the *input–process–output* thinking strategy are very valuable for selecting tools and equipment and for using them effectively and efficiently. The major difference in applying the thinking strategy to tools and equipment rather than to materials is the focus of the thinking.

When applying the thinking strategy to use *tools* and *equipment* effectively and efficiently, the primary focus is on selecting the best equipment for the application, and on equipment operation, condition, and life. However, considerable attention must also be given to *material* variables because they affect equipment condition and the ability of the equipment to achieve the desired results.

When applying the thinking strategy to use *materials* effectively and efficiently, the primary focus is on the material input, process, and output variables. Less attention is given to variables affecting *tools* or *equipment* condition.

When applying knowledge of variables and the input, process, and output thinking strategy, the impact that tools and work processes have on PEMEO must also be considered. The following table identifies factors that must be considered when selecting and using tools for production or maintenance work. Factors include:

- the desired results
- techniques to effectively use the tool
- damage to equipment and materials
- risk to health, safety, and the environment
- the effect on others and the organization

Factors for Selecting and Using Hand and Power Tools	
Factor	**Examples or explanation**
Desired result • What is the goal to be achieved through the use of the tool?	Measuring, cutting, moving, fastening, finishing.
Quality of the result • How good are the desired results?	Accuracy, fineness of finish, and time to complete the task.
Characteristics of materials or object to be worked on and tool function • What are the types, sizes, and locations of materials or objects the tool will be used on? • What tools are designed to do the specific work? • What type of tool will perform the desired action?	Specific tools and cutters are used for cutting specific materials (e.g., a wood saw blade cuts wood and plastic but does not work well for cutting most metals).

(continued)

Factors for Selecting and Using Hand and Power Tools	
Factor	**Examples or explanation**
Wear to tools • How will the tools wear as a result of use? • What can affect the life of the tools?	A grinding disk or a diamond saw blade can be used to cut concrete. The grinding disk is low cost, cuts inefficiently, and wears out quickly. The diamond blade cuts efficiently, lasts longer, but is easily damaged and costly to replace.
Damage to tools and safety hazards resulting from damage • What damage can occur to the tool during use? • What are the potential safety hazards if the tool is damaged?	A framing hammer should **not** be used in place of a ball-peen hammer to strike steel surfaces. The framing hammer's face is surface hardened where as a ball-peen hammer's face is softer. Upon impact with a steel surface, the head of the framing hammer can fragment, damaging the hammer and potentially causing personal injury.
Damage to materials • What damage can occur to the materials being worked on? • What damage can be caused to equipment components being worked on?	A ball-peen hammer should not be used in place of a soft-faced hammer to strike precision surfaces. The ball-peen hammer can damage the precision surface by causing dimples. If a ball-peen hammer is used, the precision surface must be protected (e.g., by using a soft material such as wood or brass as the strike surface). The wrong tools or the misuse of a tool can damage materials (e.g., a high speed saw can melt plastic instead of cutting the plastic, a belt sander improperly used can cause gouging and misshaping of the surface).
Damage to adjacent components, equipment, and fixtures • What is the risk of causing damage to other components, equipment, and fixtures?	Sawing or drilling through materials can result in damage to the supporting surface (e.g., workbench, sawhorse). Using a cutting torch to remove rivets can result in overheating of adjacent metal, changing the characteristics of the metal (e.g., weakening steering components on a vehicle). Heating part of a metal component can cause heat to travel to other components, damaging plastic and rubber parts. In electronics, a test probe can slip, causing a short circuit which could damage electronic parts. When grinding metal close to electronic equipment, metal dust can damage the electronics.

(continued)

Factors for Selecting and Using Hand and Power Tools	
Factor	**Examples or explanation**
Quality of materials • If the properties or characteristics of the materials change, what is the impact on other factors (e.g., wear of tools, safety)?	Changing from plywood to particleboard can dramatically reduce the life of the cutting edges on saw and jointer blades. Some particleboards contain a considerable amount of steel particles which damage the cutting components.
Efficiency of work • Is time a factor? • What tool will do the job fastest and with the least effort?	A wood surface needs sanding. A power sander is more efficient than a hand sander. With both sanders, a coarse grit sandpaper is used first for quick removal of material to eliminate surface imperfections and scratches. Progressively finer grit sand papers are used to achieve a smooth surface. Some power tools have variable speeds. The speed of the tool affects efficiency and effectiveness. For example, a drill operated at too high a speed reduces cutting effectiveness and creates excessive heat.
Effectiveness (quality) of work • Will the tool produce the quality results required? • Do tool operating speed and user technique affect the efficiency and effectiveness of the tool?	Different tools can perform the same function (e.g., cutting). However, the quality of results may vary when using different tools. Many tools operate most effectively to achieve the desired results when used in a specific way. The operator's skills in using the tools can be an important factor. The operator's direction of force, amount of force, speed, and alignment of the tool are some factors affecting the effectiveness of operation. For example, an electric jig saw makes a smooth cut when, for a specific reciprocating speed, the rate of travel does not exceed the point where the wood surface begins to splinter. The operator must apply the force in the direction of the cut; any side force results in a cut that is not square to the surface.
Health and safety • What are the risks to people while the tool is being used?	Power hand tools often have moving parts that can cause personal injury. The type of tool and its application can affect the health and safety of personnel in many different ways, including impact injury and exposure to dust and noise. (*SafeThink* provides detailed information on agents of cause and control.) Availability of safety equipment is a factor. For example, a gas cutting torch may be selected. If the proper safety glasses are not available, a power hacksaw would be a better choice.

(continued)

Factors for Selecting and Using Hand and Power Tools	
Factor	**Examples or explanation**
Environment • What are potential impacts to the environment?	Oils washed from floors in an automotive repair shop contain heavy metals. The oil, contaminants, and water must be directed to a sump for collection and disposal. Used oils and solvents must be disposed in the designated containers. Used materials can often be recycled and used elsewhere.
Organization • How does the work affect others?	Often work must be coordinated with other workers so that the work does not interfere with others doing work and put them at risk of injury. (*JobThink* and *MetaThink* provide additional information on working effectively with others.)
• Is the user qualified to use the tools safely and effectively?	For many tools, formal training is required to ensure safe and effective use.

Note that in the table, safety and environment are towards the end of the list. Often the equipment, materials, work processes, and surroundings must be known to effectively identify safety and environmental concerns.

The *quality* of tools also affects many of the criteria for optimizing your performance and controlling losses, including:
- tool life
- efficiency of work
- effectiveness of results
- damage to equipment components
- damage to materials
- health and safety
- tool maintenance
- tool costs/production costs

NOTE

Often workers debate which manufacturer produces the best tools for the job. An important consideration is to select tools that enable you to do your work safely and efficiently, are comfortable to hold, and are not tiring to use. Because body shapes and sizes vary from one person to the next, a tool right for someone else may not be right for you. If you experience discomfort when using a tool, consider using a different brand or model.

When looking for information on the web to compare tools, try using the term *ergonomics* in your searches.

Maintenance of hand and power tools may be limited to:
- cleaning
- sharpening or replacing cutting components
- lubricating moving parts
- aligning or adjusting components

When hand and power tools become worn or damaged, the tools may sometimes be replaced instead of being repaired because replacement costs may be similar or lower than repair costs.

LEARNING ACTIVITY 1

Select and use tools effectively

This learning activity helps you refine your thinking skills to use the *input-process-output* thinking strategy, *variables*, and *PEMEO* to:
- determine what is important to your job
- define the quality of results
- select tools and equipment

1.   Identify a work or technical process and answer the following questions. Examples of work processes are: cutting the lawn, cutting sheets of plywood using a table saw, removing snow from a sidewalk or parking lot, welding metal components. Examples of technical processes include using stationary equipment that refines materials (e.g., screening gravel for pea size rocks) or manufacturing a product (e.g., windows, plastic parts).

The work or technical process is: _____

_____

2. Fill in the table below for a specific job function (e.g., operator, assembler, maintenance person). Not all blanks need to be filled in.

Job function: _____

Variable	Units of Measurement	Static or Dynamic	Controllable or Non-Controllable
Key *input* variables that must be changed			
Key *input* variables that must **not** be changed			
Process variables that create the change			
Process variables that change			
Output variables that define the quality of results			

3.   Think of a *work* process such as preparing a food item or cutting plywood to make cabinet drawers.

The *work* process is: _____

Using the *output–input–process* thinking strategy, select the best tool or equipment to carry out one action or task.

Output variables that define the quality of results:

_____

_____

_____

Input variables that must be changed:

_____

_____

_____

Input variables that must **not** be changed:

_____

_____

_____

Several different tools or equipment may be used to make the change. List the tools and equipment that could be used and the process variable that makes the change (e.g., cutting, grinding, slicing). Identify the strengths and weaknesses of each tool in achieving the desired quality of results (i.e., affects the results, and the undesirable results).

Tool or Equipment	Variable that Makes the Change	Affects the Quality of Results
		_____ _____ _____ _____ _____

(continued)

Tool or Equipment	Variable that Makes the Change	Affects the Quality of Results
		_____ _____ _____ _____ _____
		_____ _____ _____ _____ _____

The best tool or equipment to achieve the desired quality of results is:

_____

Quality of results may not be the only criteria or factor to consider when selecting the best tool or equipment for the job. Using the previous table, select the tool that best meets the following criteria:

The fastest tool or equipment for doing the work is:

_____

The safest tool or equipment for doing the work is:

_____

4. The work process can have adverse effects on PEMEO. Using the work process and desired tool or equipment from question 3, fill in the following table. Refer to the table *Factors for Selecting and Using Hand and Power Tools* for examples.

Tool/Equipment: _____

Factor	Effect on PEMEO
Desired result	
Quality of result	
Wear to tools or equipment	
Damage to tools and safety hazards resulting from damage	
Damage to materials	
Damage to adjacent components, equipment, and fixtures	
Quality of materials	
Efficiency of work	
Effectiveness (quality) of work	
Health and safety	
Environment	
Organization*	

see following page

> *Below is a list of some conditions, actions, and events that apply to the organization domain.
>
> ☐ change in the work process
>
> ☐ shortage of staff
>
> ☐ lack of competent workers assigned to a specific task
>
> ☐ work group introduces new hazard
>
> ☐ third party fails to maintain equipment/facilities
>
> ☐ supplier changes standards or composition of components or materials
>
> ☐ unpredicted large customer order
>
> ☐ cancellation or delay of large customer order (e.g., may have excessive inventory on site)
>
> ☐ failure to communicate priorities
>
> ☐ lack of documentation
>
> ☐ administrative process inadequate for maintaining inventory
>
> ☐ failure to carry out routine safety or quality inspections
>
> ☐ failure to follow up on identified safety or quality deficiencies

5. On the previous table, check off the conditions, actions, or events originating from the organizational domain that affect your job and work group.

6. On the Job Aid list of *Generic Questions for Using Tools and Equipment Effectively and Efficiently*, check off the questions that apply to your work and equipment.

Mobile and Stationary Equipment

Mobile and stationary equipment can be portable, movable, mobile, or fixed. Examples of mobile and movable equipment are: portable electric generators, portable heaters, garden tractors, trucks, graders, movable gravel crushers. Examples of fixed equipment are: office equipment, and fixed processing and manufacturing equipment such as gas compressors, pumps, sanders, and wood fiber refiners. Mobile equipment moves to transfer material; stationary equipment is fixed in one position and, generally, material is fed into the equipment. There is no clear distinction between the two categories. For example, a tower crane is fixed in one position (stationary) at a building site and moves materials for construction. However, the general thinking strategies to operate mobile and stationary equipment safely and effectively apply equally to both types of equipment. The key distinction between mobile and stationary equipment is that the types of variable can be different.

Mobile and stationary equipment tend to be more technically complex than power tools. All of the optimization and loss control factors that apply to hand and power tools can also apply to mobile and stationary equipment. However, when working with mobile and stationary equipment, there is a greater emphasis on extending equipment life and maintaining equipment reliability and performance.

Over the life of equipment, reliability and performance may decrease, potentially increasing the need for additional maintenance. Damaged equipment could result in work slowdowns or stoppages. Damage to stationary equipment in a process or production line can sometimes cause the entire system to shut down. Damaged and worn equipment is often repaired because repair costs are usually lower than replacement costs.

Mobile equipment often has a limited number of functions which can be used in a variety of applications. For example, a crawler tractor can be used to pull equipment, dig house foundations, load trucks, and level ground.

If the equipment is used for an application for which it was not designed, the equipment could experience premature wear or damage. For example:

- In board manufacturing plants, a front end loader is equipped with a large chip bucket to move wood chips and sawdust. This bucket can be detached so that a smaller bucket (for moving heavier sand, gravel, etc.) can be installed. Using the chip bucket to move sand or gravel can damage the hoisting mechanism/hydraulics and decrease the loader's lifting stability.
- A half-ton truck is designed to carry a variety of dry loads. If the truck is used to carry liquid concrete, the surface of the box could be damaged from abrasion and corrosion. The concrete could also harden onto the surface of the box. The truck could become overloaded and unstable when driven.

Stationary or fixed equipment usually has only one function and one application. For example:

- gravel crushers reduce large rocks to smaller-sized rocks
- gas compressors increase the pressure of gas supplied from a limited number of sources
- sanders in a board manufacturing plant sand boards made in the production line

Stationary equipment is often combined with other equipment to form a process system or production line. The rate and/or capacity of any piece of equipment in the system can affect the rate or capacity of the entire system. If the capacity of any piece of equipment is small in relationship to the rest of the equipment in the system, a bottleneck is created. If a piece of equipment functions abnormally and affects the quality of results, other equipment and processes can be affected. If any piece of equipment fails, the system may have to be shut down.

There are many similarities between mobile and stationary equipment regarding operating and maintenance issues. However, there are enough differences in variables that this book addresses mobile and stationary equipment separately.

EquipThink™

Mobile Equipment

Mobile equipment includes trucks, graders, frontend loaders, forklifts, and many cranes. The input, process, and output thinking strategy can be very useful in selecting the *best* type of equipment to do the job. Depending on circumstances, either one of the following thinking sequences can be used:

- *input–output–process*
- *output–input–process*

4.1 Select Mobile Equipment

To make a satisfactory selection, you need to understand both the output and input variables. When working with material, output variables that define the quality of results may include dimensions and composition. Minimizing damage is often an important criterion for quality of results when transporting or moving materials.

Input variables that must be considered relate to properties and characteristics such as those identified in previous sections of this book. Access to materials may also be important. For example, it may be difficult to access the work area with mobile equipment when landscaping a residential home. Size and weight are often variables that must be considered when moving and transporting materials and goods.

Having identified the desired results and the input issues, consideration can be given to the process and equipment that can be used to do the work. Equipment function and size are often important variables. Often, several different types or models of equipment can be used to do the work. For example, backhoes, crawler tractors, and frontend loaders are all earth-moving equipment. However, for a specific job, one type of equipment may be more useful and effective than other types.

Undersized equipment can reduce the work efficiency and the quality of results. Undersized equipment can also experience excess stress, affecting the equipment condition and life. Overloaded equipment can also place people and the environment at risk. On the other hand, oversized equipment may make it difficult to access specific areas and may increase operating costs.

Refer to *MatThink* for more examples of selecting the most suitable equipment for the job.

4.2 Operating Mobile Equipment Effectively

After selecting the equipment, you need to plan the work and operate the equipment in a way that:
- achieves the desired results (effective work)
- minimizes time, effort, and energy (efficient work)
- minimizes wear and damage to equipment (i.e., extend equipment life and reliability)
- minimizes material damage, waste, and costs
- minimizes harm to people and the environment

Work processes require you to have the skills to operate the equipment safely and effectively. However, to work efficiently, you need to plan your work and remain vigilant about the critical variables that could cause damage to equipment, materials, and surrounding objects and structures.

The *output–input–process* thinking strategy can be very useful for planning your work.

Output

First, identify the output variables that define the quality of results. For example, when moving earth, elevation, dimensions, and location are important variables defining the quality of results. When transporting materials and products, minimizing damage to the load, delivery location, and arrival time may all be important variables. This knowledge about the input and work process variables is useful for you to determine the actions needed to achieve the desired results.

Input

Next, examine the input variables. Determine the material variables that must change and those that must not change. When moving earth, such as digging holes, grading, and loading trucks, you need to also consider the layout of the site. Other equipment and structures may require you to plan your work so that you:

- have room to work (e.g., when digging a trench to/from a structure, dig from the structure outwards)
- place the earth in a location easy for trucks to access and be loaded
- prevent damage to adjacent structures
- prevent injury from contacting overhead electrical power lines
- prevent injury from contacting buried utilities such as electrical and gas lines

Input material properties and characteristics must be considered when planning work to achieve the desired results. An example is using mobile equipment to dig a basement. When moving materials that are loose and not compacted (e.g., digging in sand) the sand can cave in. The top of the hole must be wider than the bottom. The walls of the sand must be sloped enough to prevent movement. An efficient work strategy must be determined to remove the earth so that the walls are sloped and the hole is to the proper elevation, dimensions, and location.

Following are two possible work planning scenarios for excavating using a crawler tractor:

1. *For a shallow excavation, the excavation could be slightly smaller than the finished dimensions and then the walls sized and sloped to the desired results.*

2. *For a deep excavation, the slope must be determined before starting work. The top dimensions of the excavation are considerably larger than the bottom dimensions. The walls would be sloped as the earth is removed.*

Try this:

You are digging a trench that you consider to be neither shallow nor deep. What variables would you take into account to determine the best work strategy?

_____

_____

A change in input material properties and characteristics may require you to change the work strategy and operation of the equipment. For example, in a deep excavation, the soil structure could change from sand to clay, mud, or gravel. Soil testing (drilling boreholes) may be done before excavation to help plan the excavation process.

When transporting products, a major concern is to **not** change the properties and characteristics of the products. Damage from handling, vibration, or shifting of loads must be minimized. Food products are often refrigerated. Failure of a refrigeration system could result in the product losing value due to spoilage.

Knowing the type of load and the critical variables that must not be affected are important for planning to be able to transport the product safely without damage to the load and putting PEMEO at risk. Planning the route may also be important to ensure that height and load restrictions are not compromised. When traveling in remote and unpopulated areas, there is always the possibility of being stranded. Planning should include ensuring that roadside warning markers, communications, and personal survival equipment are available.

When planning, identify the critical variables for output, input, and process that must be considered.

Try this:

Select a task that you routinely carry out. Identify **two** variables each for output, input, and process that you must consider when planning your work.

The task is: _____

Output variables (2): _____

Input variables (2):_____

Process variables (2): _____

Work Process

Using mobile equipment is a work process because the focus is on doing various types of work or jobs. Work processes using mobile equipment often require the operator to perform the following tasks:
- plan the work
- carry out pre-work checks of the equipment (and possibly the site)
- shut down the equipment
- start up the equipment
- hook up towed equipment
- operate the equipment
- record equipment conditions and problems

Plan the work

The first part of planning may take place away from the equipment and worksite and involve understanding the output and input variables. When the equipment, worksite, or load are closely examined, the plan may have to be refined. While working, if there is a change in input materials, site/road conditions, or the desired results, the work plan may have to be revised. Communication (and, sometimes, documentation) is required when changes occur and/or the plan is revised.

Shut down the equipment

When learning to operate equipment, one of the first things you should learn is to shut down the equipment. Being able to shut down the equipment is important should things go wrong. Often organizations have policies about shutting down mobile equipment to ensure the equipment cannot harm others and is secure. Examples of policies for shutting down equipment include:

- refueling
- locating and orientating the parked equipment (e.g., backing into position)
- preventing the equipment from rolling (engaging the breaks, chocking, blocking)
- positioning moving components (e.g., loader bucket) so that they will not harm anyone should the components suddenly move
- parking equipment on job sites so that the equipment does not get damaged or interfere with access and movement of other equipment
- allowing the prime mover to cool down before shutting it off if it has been working hard
- preventing natural elements from causing damage (e.g., preventing rain water from accumulating in a truck box)
- securing the equipment and load to prevent theft
- preventing the equipment from starting during maintenance
- documenting run time, damage, and equipment operating performance and condition

Carry out pre-work checks

Before starting work, walk around the equipment to ensure that it is in good operating condition and safe to use. Equipment that is towed must also be checked. The checks fall into the categories listed in the following table. Knowing the categories can help you to be more thorough in checking the equipment when you do not have a check sheet. The categories of checks can also help you to think through the required checks when working with equipment new to you.

Check	Example
damage	to body, tires, lights, mirrors, windows, and fluid leaks
fluid levels	of engine, auxiliaries, windshield washer
position	of seat, mirrors, components (moving parts that perform the work)
gauges after startup	indicating the status of engine and auxiliary variables such as oil pressure
operation of safety devices	lights, horn, backup alarm, brakes
operation of components	moving parts that perform the work
clearances	overhead lines, closeness of structures and people

Sometimes, as part of the pre-start checks and/or startup, equipment has to be adjusted for a specific application (e.g., correct size of auger is available or attached to drill piles).

Start up equipment

Before starting, check to ensure that no powered components are engaged that would make the equipment lurch when started. Follow the startup procedure for the equipment. When the equipment starts and is warming up, check the gauges to ensure normal operation. In addition to using gauges to receive information about the equipment operation, use your senses to:

- listen for unusual or abnormal noises (i.e., noises that are different from the noises when operating the equipment at other times)
- feel the vibration to determine if it is different from previous operation
- look at the operating parts to ensure they are functioning as expected
- note any unusual odors

If you sense anything that is abnormal or different, investigate further. If the equipment operation is very abnormal (e.g., making an unusual or loud banging noise), shut down the equipment.

If practical, check the moving components to ensure they function as expected, especially after maintenance has been performed.

Hook up towed equipment

Some equipment is not self-propelled and must be towed by a truck or tractor.

- Make sure that the equipment being hooked up is blocked so it cannot move.
- Before backing up, ensure that others do **not** stand between the towing equipment or behind the equipment to be towed. Check the proximity of structures and other obstacles.
- Make sure the tractor's or truck's brakes are on before leaving the vehicle to prevent rolling. If the vehicle is parked on a slope, block the wheels.
- If the hitch must be lowered or raised for alignment, keep your hands clear of pinch points and your feet clear of the fall area in case the hitch or towbar drops unexpectedly.
- If two people are hooking up the equipment, make eye contact and communicate clearly about the impending actions. Make sure any secondary safety devices (e.g., safety clips, chains) are hooked up.
- Some attached equipment needs to be powered from the power takeoff of a tractor. Remove loose-fitting clothing, scarves, jewelry, dangling jacket cords, and other items that can become caught on the rotating shaft. Tie long hair back. **Never step over** the power takeoff shaft to check or adjust equipment.

Before starting out, check the safety devices:

- Head, tail, and signal lights on both the vehicle and towed equipment are clean and work properly.
- Temporary lights on the towed vehicle are clean and work properly.
- On wide farm equipment (e.g., cultivators and sprayers), the fold supports are secure.

- Warning signs (e.g., slow-moving vehicle placard, clearance flags) are secure.
- Audible and visible backup signals work properly.
- Others are clear of the moving zone

Operate the equipment

Licensing, regulations, and company policies ensure that minimum operating standards and operator qualifications are met. The primary goals are to operate mobile equipment safely, effectively, and efficiently. To be an exemplary performer, you must continually work at refining your operating abilities. You need to continually pay attention to:

- the desired results
- safety
- equipment performance and condition
- your skills in operating the equipment

Always keep the desired results in mind. Think of the criteria for quality and how your work strategy and operating skills can best be used to achieve the desired results.

Pay close attention to your work surroundings. Watch for stationary objects and for equipment and people entering your work area. A change in weather conditions can reduce visibility and make surfaces slippery or muddy. In mountainous areas, falling rocks and slides can be a hazard. When repairing roads, traffic may flow very close to your equipment.

How you use the equipment and equipment performance and condition can compromise safety. Follow recommended operating practices to ensure the equipment does not tip. Avoid excessive speed that can cause a loss of control and reduce the response time needed to react to people and equipment entering your work area.

Poor equipment condition and performance (e.g., poor brakes, low engine power) can make it difficult to achieve the desired results and potentially put people at risk. Always monitor the gauges, use your senses to monitor operation, and continually be alert for changes in equipment performance.

The way you operate the equipment can make a major difference to equipment condition and life. Operate the equipment within the recommended ranges for variables. For example, do not lug an engine at low rpm or operate at excessive rpm. Where practical, minimize stress to the equipment. For example, a short braking distance requires hard braking which reduces brake life and may cause skidding. Increasing braking distance and braking slowly reduces brake wear and gives you more control of the equipment.

Impact shock is a frequent cause of equipment damage and premature wear. Impact shock is caused when excessive force is applied over a very short time. For example:

- driving over pot holes at highway speeds causes excess stress on the vehicle's steering and suspension
- jerking on a tow line or cable. For example, when using a rope or cable to pull another piece of equipment or lift a load, first slowly take up the slack and then increase the force. Rapidly taking up the slack so that the cable snaps tight causes excessive stress (impact shock) on the connecting components and cable. The connecting components or cable can fail, causing the cable to rapidly snake through the air, potentially striking people or objects in the vicinity.
- making rapid changes in the direction of movement of functioning components (e.g., back hoe). With an engine operating at high rpm, any rapid change in operating the equipment causes excessive stress on the components, reducing component life and potentially causing component failure.
- dropping a heavy load onto a truck bed causes excessive force on the truck bed and suspension
- a grader blade hitting a protruding man-hole cover causes a shock to the grader and can potentially break the blade (especially in cold weather)

Changes to the properties and characteristics of input material can affect the equipment condition and quality of results. You may have to adjust the way you work because a change in materials may result in the materials responding differently to your operational moves. Changes to materials can also affect your safety. For example, a change from hard-packed soil to loose soil could increase the potential for equipment to not respond as quickly to your operational moves (e.g., turning when moving the equipment) and/or increase the likelihood of tipping.

To be an exemplary equipment operator, you need to:
- have a clear understanding of the criteria that defines the quality of results (output)
- understand the input variables and how they affect the work and equipment performance and condition
- understand how the equipment operates
- know the capabilities and limitations of the specific equipment, size being one important variable
- have the skills to operate the equipment effectively (achieving desired results) and efficiently (minimum effort, energy, and time without harm to PEMEO)

NOTE

Sometimes working efficiently gets the work done quickly, safely, and with quality results but can cause added stress to equipment, increased fuel consumption, and increased pollution.

When learning to operate a specific piece of equipment, you need criteria to judge your own performance or skill. Terms such as *smoothly, accurately, with precision* are often used to define exemplary skills. However, you need to observe equipment being operated smoothly, accurately, or with precision to understand the meaning of those terms when applied to specific skills. When you know what to look for when assessing your own operating skills, you can be more effective at developing your skills. Keep in mind that skilled operators also minimize stress on equipment and prevent damage to PEMEO—the most important being ensuring their own safety and the safety of others.

To operate specific equipment safely and effectively, government regulations or your own employer may require you to take a specific training course (e.g., forklift training).

During and after taking the course, think about the output, input, and work process variables. Wherever possible, observe others operating the equipment to learn and refine your skills. Talk to others about the equipment and operating issues. Asking the right types of question also increases your ability and flexibility to learn to operate different types of equipment. The Job Aid for this book lists the most important questions to ask yourself or others. To be exemplary, you must have a commitment to be the best you can be and practice, practice, practice. To maintain your exemplary skills, you must continue to practice and learn from others.

Record equipment conditions and problems

Poor equipment condition can put you and others at risk of injury. Immediately report any equipment condition or problem that could affect safety. You may have to shut down and secure the equipment until the condition is corrected (e.g., brake failure).

Some jurisdictions and/or companies require that a log book (or work repair order) be used to record equipment deficiencies. The reported deficiencies must be corrected in a reasonable time period.

LEARNING ACTIVITY 2

Use mobile equipment effectively

This learning activity helps you refine your input, process, and output thinking strategy and the use of variables to determine what is important to operating equipment. The learning activity also gives you the opportunity to apply a structured thinking strategy to ask useful *What if... ?* questions.

1. Identify a work process (task) using mobile equipment and answer the following questions. Examples of using mobile equipment are:
    - a tractor-pulled mower to cut grass in parks
    - a front-end loader to remove snow from parking lots
    - a mobile welder to weld metal components in the field
    - a rock screener to separate gravel for pea size rocks

- a grader to level a road
- a tractor-trailer to transport consumer goods

The work process is: _____

The equipment used is: _____

1a. What are the critical *output* variables (quality, quantity, timeliness)?

_____

_____

_____

1b. What are the critical input *variables* (e.g., material properties and characteristics, surroundings, weather conditions)?

_____

_____

_____

_____

1c. What are the critical *process* variables (e.g., pushing, moving, hauling, digging, leveling, loading, crushing, lifting)?

_____

_____

1d. What actions or conditions can cause undesirable results?

_____

_____

1e. What changes to the inputs can cause you to change the way you carry out the work?

_____

_____

1f. How can an operator cause premature wear and damage to the equipment (i.e., reduce the life and condition of the equipment)?

_____

_____

_____

1g. There is always a possibility of harm being caused to PEMEO. Fill out the following table. For each category of PEMEO, identify two conditions, actions, and/ or events that could cause an incident. Identify the immediate effects. Note that using the PEMEO strategy is very useful for identifying *What if... ?* questions.

What can go Wrong with PEMEO?		
Domain	**What if... ?** *(conditions, actions, events)*	**Immediate Effect**
People	• What if	•
	• What if	•
	• What if	•
	• What if	•
Equipment	• What if	•
	• What if	•
	• What if	•
	• What if	•
Materials	• What if	•
	• What if	•
	• What if	•
	• What if	•
Environment	• What if	•
	• What if	•
	• What if	•
	• What if	•

(continued)

What can go Wrong with PEMEO?		
Domain	**What if... ?** *(conditions, actions, events)*	**Immediate Effect**
Organization	• What if • What if	• •
	• What if • What if	• •

2. Use the work process and mobile equipment you listed in question 1. In the table below, list three critical variables you pay special attention to when performing each step of the task.

Action Steps	Critical Variables
Shut down equipment	• • •
Plan the work	• • •
Perform pre-start checks	• • •
Start equipment	• • •
Hook up towed equipment	• • •
Operate equipment (critical operating skills)	• • •
Record equipment conditions and problems	• • •

3. Give an example where you have observed equipment being overloaded or overworked. Identify the safety issues and impact on the equipment.

_____

_____

4. In the Job Aid, there are two lists that apply to mobile equipment:
   - *Critical Thinking Questions about Work Processes Using Mobile Equipment*
   - *Critical Thinking Questions about Operating Mobile Equipment*

   Check off the questions that apply to your work and equipment. Add any questions that are work or equipment-specific.

Stationary Equipment

There are many types of stationary equipment including offset presses, milling machines, and oil, gas, and petrochemical vessels, compressors, and pumps. Stationary equipment often has only one function and one application. For example, a gas compressor increases the pressure of gas supplied from a limited number of sources; a sander in a board manufacturing plant sands boards made in the production line. Fixed equipment is often combined with other equipment to form a process system or production line.

5.1 Operate Stationary Equipment Effectively

To be exemplary at operating equipment, you must:
- have a clear understanding of the criteria that defines the quality of results (output)
- understand the input variables and how they affect the work and equipment performance and condition
- understand how the equipment operates
- know the capabilities and limitations of the specific equipment, size being one important variable
- have knowledge about equipment operation and control
- carry out procedures effectively, efficiently, and safely

- make adjustments to operating equipment in response to changes in input variables or changes in desired outputs
- recognize and respond to abnormal operation

Operating stationary equipment effectively and efficiently with minimum stress on the equipment often requires considerable knowledge about variables. Within your roles and responsibilities, you must not only identify the critical output, input, and process variables, but also know the normal settings or operating ranges for the variables. You must also know the **reasons** for operating variables to be maintained at specific settings or within given ranges.

5.2 Output-Input-Process Thinking Strategy

Generally, the operation of stationary equipment involves more mental skills and less physical skills than the operation of mobile equipment. The purpose or function of most stationary equipment is to *change* materials to produce a specific output.

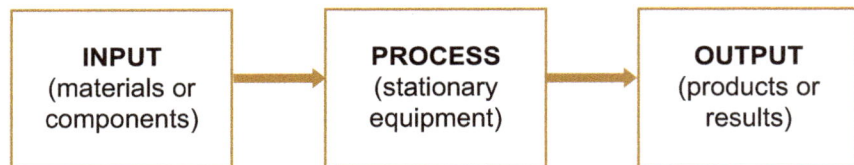

INPUT (materials or components)	→	PROCESS (stationary equipment)	→	OUTPUT (products or results)

Examples of stationary equipment making changes to materials include:
- milling machine shaping a component
- extruder producing plastic rods
- blast furnace heating iron ore
- punch press making holes in metal plates
- gas oven baking breakfast cereals in a production line
- mixer blending ingredients for pizza dough
- amine vessel (tower) removing H_2S from natural gas
- centrifugal pumps and valves directing product to bulk hydrocarbon storage tanks

The following illustrations show some common combinations of inputs, processes, and outputs.

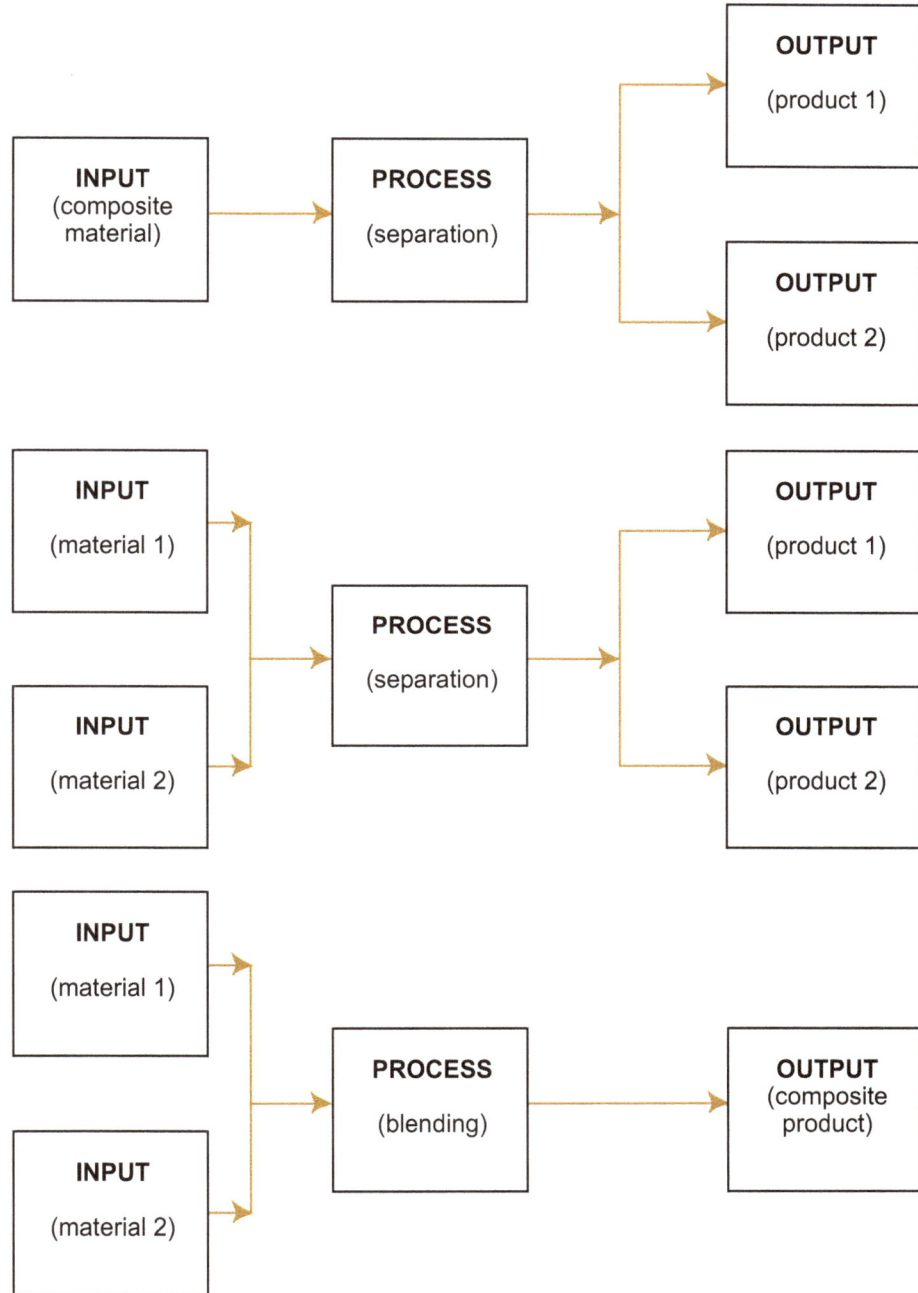

```
┌──────────────┐        ┌──────────────┐                    ┌──────────────┐
│    INPUT     │        │   PROCESS    │        ┌──────────▶ │    OUTPUT    │
│ (composite   │──────▶ │ (separation) │──────▶ │            │ (product 1)  │
│  material)   │        │              │        │            └──────────────┘
└──────────────┘        └──────────────┘        │            ┌──────────────┐
                                                └──────────▶ │    OUTPUT    │
                                                             │ (product 2)  │
                                                             └──────────────┘
```

```
┌──────────────┐                                            ┌──────────────┐
│    INPUT     │                                 ┌────────▶ │    OUTPUT    │
│ (material 1) │─────┐                            │          │ (product 1)  │
└──────────────┘     │   ┌──────────────┐         │          └──────────────┘
                     ├─▶ │   PROCESS    │──────▶ │
┌──────────────┐     │   │ (separation) │         │          ┌──────────────┐
│    INPUT     │─────┘   └──────────────┘         └────────▶ │    OUTPUT    │
│ (material 2) │                                             │ (product 2)  │
└──────────────┘                                             └──────────────┘
```

```
┌──────────────┐
│    INPUT     │
│ (material 1) │─────┐
└──────────────┘     │   ┌──────────────┐                    ┌──────────────┐
                     ├─▶ │   PROCESS    │──────────────────▶ │    OUTPUT    │
┌──────────────┐     │   │ (blending)   │                    │ (composite   │
│    INPUT     │─────┘   └──────────────┘                    │  product)    │
│ (material 2) │                                             └──────────────┘
└──────────────┘
```

The following illustration shows a specific example of applying the input, process, output concept.

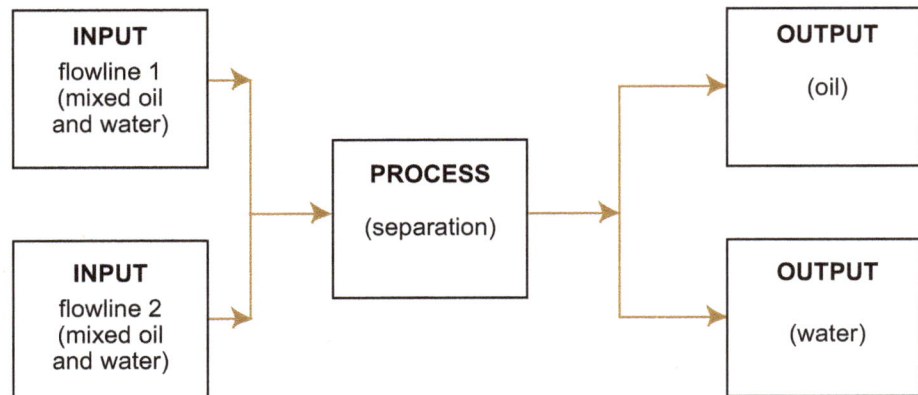

The *output-input-process* thinking strategy is very useful when working with stationary equipment, especially manufacturing and technical processes. Specifications for material and equipment variables are stated as a:

- **fixed** value (e.g., 180°C, 750 rpm, 1250 cm)
- **tolerance** (e.g., 3.5 cm + 0.1 cm, 10.00 kg + 0.25 kg)
- **maximum or minimum** value (e.g., no more than 0.5% water, no less than 4% butane)
- **range** (e.g., between 600 and 900 kPa, between 4.5 and 5 cm)

Critical variables are those variables that relate directly to your job and:

- define the quality of results
- affect the quality of results
- define the quality of input materials
- create change
- are changed
- can cause harm to equipment and other PEMEO domains
- are affected by PEMEO
- are important to operating equipment effectively

Output Variables—usually the output variables that define the quality of results are fixed. However, in manufacturing the output variables may be fixed for each production run, but the desired quality of results may be different for each run. Desired results often fall into three categories:

- quality of products
- rate (quantity) of output
- timing/timeliness

Input Variables—the properties and characteristics of input materials can affect the quality of results and equipment condition. First, identify the variables that must be changed and those that must **not** be changed. Next, determine the input capacity of the equipment. Manufacturers rate the input material capacity of their equipment to ensure that:

- the equipment is not damaged
- the equipment's life is not shortened
- the results are satisfactory
- safety and the environment are not put at risk

Examples of input capacity variables, include:

- maximum load of 2 tonnes
- maximum size of steel plate is 10 mm
- minimum thickness of plywood sheets is 3 mm
- maximum suction pressure of 200 kPa
- maximum diameter of rock is 25 cm
- maximum temperature of 120°C
- minimum length of material is 50 cm

The following diagram illustrates the operating limits of a variable and the ideal operating setpoint.

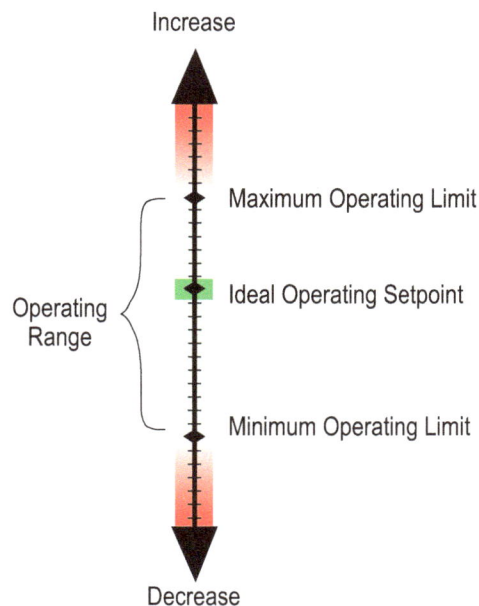

Operating Limits and Setpoint

If the input materials cause the equipment to operate outside of the minimum or maximum recommended limits, the equipment could experience premature wear, damage, or failure. For example:

- increased suction pressure on a gas compressor increases the load and strain on the compressor
- an increase of granite in kimberlite makes the ore grinding process more difficult
- an increase in water content in a gas stream increases the load on dryers and dehydration systems in gas plants
- an excessive load on a diesel engine can dramatically shorten the life of the engine
- a very low liquid flow rate to a pump can cause the pump to overheat and/or cavitate

The following illustration shows a person using the wrong criteria to determine input capacity.

Foreign material can cause harm to equipment, degrade the quality of results, or put people's safety at risk. For example:

- solid debris, solvents, and corrosive products in water can damage a pump
- rocks mixed in with vegetables that are being processed can damage cutting blades

- a slug of liquid condensate in a gas line can damage a natural gas compressor
- steel objects in wood chips can damage a fiber refiner
- H_2S (sour gas) introduced into a sweet gas stream can corrode equipment, affect the equipment's operation, and pose a safety hazard
- an increase of suspended material and plant matter in river water in spring time affects the water purification processes

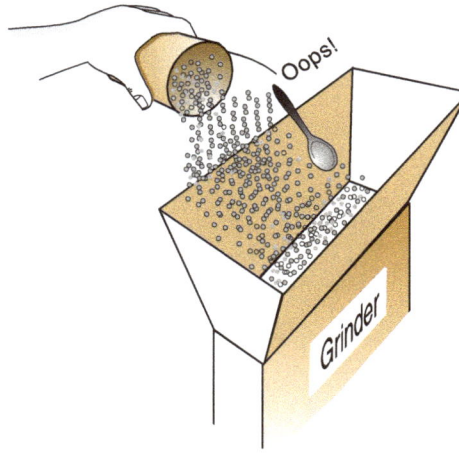

NOTE

Sometimes the stationary equipment can introduce foreign materials that can affect the quality of results (e.g., oil leaks, broken components).

Seasonal conditions can also affect the input variables and equipment operation. Some equipment needs to be operated differently in winter than in summer because the change in temperature can affect the characteristics of materials and also equipment operation. For example, the procedures to start a diesel engine or start a pump to transfer viscous liquids are different for cold and hot weather conditions.

Process Variables—There are two categories of process variables that you must pay attention to: material variables and equipment variables.

Material variables—You must monitor material variables entering and leaving the equipment to prevent harm to the equipment. You may have to adjust equipment should the material variables change.

Equipment variables—You must determine the equipment variables that create change to the materials so that you can operate the equipment effectively. You must also monitor the variables associated with equipment operation and condition and make adjustments if the variables shift. For example:

- oil filters become clogged
- equipment overheats
- equipment output drops due to wear
- equipment vibration increases
- equipment rpm increases or decreases unexpectedly

Here are the key questions to ask yourself about *output-input-process* variables about how stationary equipment operates.

Output Variables

- *What output variables define quality of results?*
- *How are the output variables measured?*
- *What are the targets or optimal ranges for output variables?*
- *Can the desired quality of results change?*
- *What is the impact of the outputs on PEMEO?*
- *What is the impact of PEMEO on the outputs?*

Input Variables

- *What are the input variables?*
- *What input variables are changed by the process?*
- *What input variables must not be changed?*
- *What input variables are static? Dynamic?*
- *What input variables are controllable? Non-controllable?*
- *How are the input variables measured?*
- *What are the targets or optimal ranges for input variables?*
- *Can the input materials contain contaminants that could affect the equipment and desired results?*
- *What is the impact of input materials on PEMEO?*
- *What is the impact of PEMEO on input materials?*

Process Variables

- *What process variables change the materials?*
- *How are the process variables that change the materials measured?*
- *What process variables that change the materials are static? Dynamic?*
- *What process variables that change the materials are controllable? Non-controllable?*
- *What process variables can downgrade the results?*
- *What is the impact of process variables on PEMEO?*
- *What is the impact of PEMEO on the process?*

Change in Variables

- *What are the indications that the process variable has changed?*
- *What are the potential consequences if a process variable changes?*
- *What do I do if the process variables change?*
- *How much time do I have to respond effectively to a change in a variable?*
- *If the input variables change, what changes do I have to make to the process variables?*
- *If the quality standards for results change, what changes to the input or process variables do I have to make?*

Operating equipment efficiently and effectively and with minimum stress on the equipment requires considerable knowledge about variables. Within your roles and responsibilities you must:
- identify the critical output, input, and process variables
- determine their normal specifications and the **reasons** for operating variables to be set at specific settings or maintained within given ranges

A basic knowledge about the equipment design and operation is important to operating safely and effectively. Equipment that performs the same function may operate very differently. For example, both reciprocating and centrifugal gas compressors compress gas. However, their design and the way they operate are very different. The operation and monitoring requirements of the two types of compressor are also very different.

Sometimes equipment is modified to improve performance. The modifications affect the equipment operation and maintenance requirements. The modifications could also reduce the life of other components. For example, replacing a gasoline engine with a more powerful diesel engine puts more strain on the driven components (e.g., transmission).

Manufacturer's literature is often a good source of operating information. However, if modifications have been made to the equipment or to maintenance requirements, the manufacturer's operating and maintenance specifications and procedures may no longer be valid (e.g., a higher temperature lubricating oil may be used than recommended by the manufacturer).

NOTE

As equipment wears, the operating limits may have to be reduced to extend equipment life and prevent equipment failure.

In addition to knowing the normal setting for input, process, and output variables, you need to know whether the variables are:
- static or dynamic
- controllable or non-controllable

This knowledge helps you determine which variables could change and whether or not you have control to respond to the changes (within your roles and responsibilities).

Generally, the changes you need to pay most attention to are:
- desired changes in the quality of results
- changes in input material variables
- changes in equipment variables
- undesirable changes in the material output variables

Should variables change, you need to know:
- **indicators** that variables have changed
- whether the change is **abnormal**
- the **cause** for variables to change
- the **consequences** of changes in variables
- your **response** to changes in variables
- the amount of **time** you have to respond effectively to the change

5.3 Equipment Operating Tasks

For each operating task involving stationary equipment, monitor the input, process, and output variables to ensure that they are within acceptable operating limits. If the variables change or go outside the desired limits, you must know how to respond effectively.

Common operating tasks for stationary equipment are to:
- shut down equipment
- perform pre-start checks
- start equipment
- bring to desired operating level
- monitor and adjust equipment

Shut down equipment

Often production and process systems operate continuously, twenty-four hours a day, every day. Some production systems produce batches. After completing each batch, the equipment is shut down and carefully cleaned.

Operators spend most of their time monitoring production and equipment performance and condition. Should a need arise, operators need to know how to shut down the equipment.

There are four common types of shutdown:

Temporary shutdown—the materials are left in place with the understanding that the equipment will be started in a short while. For example, when a process line is shut down, ore remains on the conveyors, in piping, and in the process equipment. In gas processing, the gas remains pressured in the piping and equipment. Often, equipment auxiliary systems (e.g., oil circulation and cooling) are left running so the equipment can be started immediately.

Longer-term shutdown—the material supply is shut off and the materials in the system are processed. When the equipment is shut down, little material remains in the transfer and process equipment. For example, a food processing plant shuts down each time a product batch is completed. In gas processing, the equipment and piping are depressurized as part of the shutdown process.

Equipment auxiliary systems (e.g., oil circulation and cooling) are often left running for a specific length of time after the main equipment is shut down to minimize stress on the equipment and extend equipment life.

In batch processes, the equipment may have to be locked out, cleaned (and sometimes sterilized) during shutdown.

Maintenance shutdown—specific equipment, a process, or a facility is shut down for maintenance. Upon shutdown, the equipment must be made safe for maintenance personnel to carry out their work. Electrically and mechanically locking out equipment is one method of ensuring the equipment is safe to work on. If materials are hazardous (e.g., flammable, toxic, corrosive) the materials must be removed from the equipment. The equipment must also be isolated from upstream and downstream material to prevent personnel from potentially being exposed to the material.

Emergency shutdown—equipment is rapidly shut down to prevent damaging equipment and/or harming people or the environment. In gas processing, the equipment and piping are also depressurized to a flare stack to prevent gas from escaping into the workplace atmosphere.

You need to:
- know the reasons for each type of shutdown
- recognize the conditions that determine the type of shutdown required

- be able to perform each type of shutdown safely and effectively
- communicate with co-workers before, during, and after a shutdown

Each shutdown has a different procedure. However, for all shutdowns, monitor and check the equipment to:
- ensure all components have come to rest
- determine the status of monitoring and control components to ensure PEMEO are protected (e.g., gates on raw material input chutes are fully closed to prevent materials from entering the equipment or spilling)
- determine behavior of critical variables (e.g., fluid pressure in blocked-off lines remains stable)
- identify component wear or damage

Perform pre-start checks

The condition of the equipment and materials that were left in the equipment after a shutdown affects the pre-start requirements. After a temporary shutdown, a basic check of the equipment and material variables may be all that is required. After a maintenance shutdown, extended pre-start checks may be required.

Seasonal differences can also affect the type of pre-start checks that are performed (e.g., in cold weather, heat tracing to prevent freezing may have to be checked for condition and operation). Follow your documented pre-start checklists.

Although there are differences in the pre-start checks following different types of shutdown, there are common types of checks that are performed:
- equipment condition (wear, damage, leaks, fluid levels normal)
- removal of obstructions to equipment operation (e.g., maintenance tools, electrical cords, and components that could come in contact with moving parts)
- readiness of equipment (energized, availability of fuel, brakes on/off, etc.)
- operation of auxiliary equipment (e.g., oil circulation systems). For large stationary equipment, auxiliary

systems for equipment cooling and lubrication are often started before the main equipment is started.

- configuration of supply and discharge equipment (e.g., valves opened/closed, control gates opened/closed, availability of material for input, and a place for product output)
- communication with others who will be affected by the pending startup

Start equipment

Often equipment in production and process systems must be started in a specific sequence. A *cold start* may have to be performed if equipment has been down for a long time. The equipment is operated without handling material until operating parameters are within acceptable limits (e.g., rpm, temperature, vibration).

The first materials being processed may not meet output specifications. Some of the equipment and related variables affecting the materials (e.g., circulating of hot fluids) may not reach operating parameters until a certain amount of material is processed. The first materials entering a process system may be contaminated due to maintenance or cleaning activities (e.g., air, dust, cleaning fluid residues in the process equipment).

When starting equipment, you need to monitor:
- equipment condition
- equipment operating variables
- input material variables
- process variables
- output material variables

Bring to desired operating level

Some equipment and process systems must be slowly brought up to operating level (ramped up). If the equipment and material variables do not meet normal operating parameters after startup, you may have to adjust the equipment operation. If abnormal parameters persist, determine the cause(s) and take corrective action.

| NOTE | *JobThink* describes a practical problem-solving strategy for equipment. |

Following the recommended operating procedures can minimize wear and tear to equipment and extend equipment life. Operating actions that reduce equipment life often have one common effect—they cause a **rapid change** in a variable such as temperature or force. The following table provides examples of how rapid changes in variables can reduce equipment life.

Slowly changing variables can also result in damage. For example, if a tank filled with vapor cools, it can collapse. As the vapor condenses, the vessel's internal pressure will drop below the outside pressure. The vessel will collapse if the vacuum-relief valve fails to sufficiently reduce the vacuum.

Changes to one variable sometimes affect one or more other variables (e.g., changing flow rate in a line can affect pressure).

Rapid Change to Variables Reduce Equipment Life	
Variable	**Examples**
Temperature	• A large volume of hot product entering a cold pump or vessel causes localized heating and expansion of materials. • Several unsuccessful attempts to start electric motor-driven equipment can cause the electric motor to overheat. • Piping exposed to the sun can heat up, causing thermal expansion. Liquids trapped in the piping expand, causing a major increase in pressure if there is no place for the product to go (i.e., upstream and downstream valves are closed).
Pressure	• A rapid increase or decrease in pressure can stress the wall of piping and vessels due to the rapid expansion or contraction. • A sudden decrease of pressure in a pressurized plastic vessel can damage the interior walls of the vessel. Under high pressure, fluids can impregnate the plastic. A rapid decrease in pressure causes the fluids trapped in the plastic to expand and damage the plastic. Slowly decreasing the pressure allows some of the fluids trapped in the plastic to migrate back into the vessel.

(continued)

Rapid Change to Variables Reduce Equipment Life	
Variable	**Examples**
Flow	• A sudden change or stoppage of flow in a pipe by rapidly closing a downstream valve can result in "water hammer". As the fluid rapidly slows down, there is a pressure shock wave that moves along the piping and causes the pipes to shake. • A rapid decrease in flow rate by restricting downstream flow causes an increase in line pressure and stress on piping and equipment (e.g., rapidly reducing the flow of a 100 km pipeline in response to a leak). • Rapidly opening a valve that has high upstream pressure and low downstream pressure can stress piping and vessels. Excessive local forces can be created at pipe bends as the fast traveling fluid changes direction in flow. To reduce shock, a valve is cracked (opened slightly), letting the pressure slowly equalize on both sides of the valve, before opening the valve to the desired setting.
Impact Load	• A free-wheeling engine (driver) operating at high rpm suddenly engaged with the driven equipment (load) creates excess stress on the equipment and engine. Note: Some equipment components (e.g., reciprocating parts, non-slip clutches) are designed to withstand sudden increases in mechanical force.

Operating equipment in a way that shortens equipment life often increases the use of energy and consumables. For example, an overloaded internal combustion engine may:
• consume additional fuel
• consume additional lubricating oil
• increase contamination of lubricating oil
• produce additional air pollutants

Monitor and adjust equipment

After the equipment and process/production systems are operating, they must be monitored and adjusted, if necessary:
• *equipment* variables must be monitored to ensure they do not exceed acceptable operating limits to meet production requirements and prevent equipment stress, damage, or failure. For rotating equipment, variables commonly monitored include: bearing vibration, rpm, noise, fuel

pressure, leaks, and lubricant and coolant pressures and temperatures.

- *material* variables that affect equipment operation must be monitored and equipment operation adjusted to ensure efficient production and to minimize equipment damage.
- *output* material variables such as quality, quantity, and timing must be monitored to ensure the desired results are achieved.

Before making adjustments to equipment that is part of a system, always consider the impact the change could have upstream and downstream of the equipment. Ask yourself, *How will this change affect upstream and downstream equipment operations and people?*

5.4 Monitor and Adjust Equipment

Equipment is monitored and adjusted both manually and automatically. Often production and process facilities use protective devices, instrumentation, and control systems to make the production processes safe, effective, and efficient, and to protect equipment from damage.

Protective Devices

Some equipment has protective devices that prevent the equipment from operating outside specific limits. Protective devices can function several ways, including:

- preventing equipment from starting if process variables are not within the required limits (e.g., a centrifugal pump cannot be started unless the suction pressure is above a safe minimum pressure)
- adjusting equipment operation to prevent equipment from operating outside specific operating limits (e.g., a governor prevents a gas turbine from overspeeding)
- diverting materials to reduce excessive stress on equipment (e.g., if the maximum gas pressure in a vessel is exceeded, a valve opens to vent gas to atmosphere or a flare stack)
- shutting equipment down if process operating limits are exceeded (a centrifugal pump shuts down if suction pressure drops below the safe minimum pressure)

- shutting equipment down if equipment components function outside of their design limits (e.g., if a bearing vibrates in excess of preset limits, the equipment shuts down)
- designing equipment so that, if equipment experiences excessive impact forces, minor components fail first to prevent damage to major components. For example, a sheer pin on a small engine flywheel will sheer off if the equipment experiences a large mechanical impact load
- shutting equipment down if there is a safety hazard (e.g., flammable gas is detected in the atmosphere)

Audible and visual alarms may also be activated when an operating variable approaches, exceeds, or falls below the operating limit. The following illustration is one example of alarm and automated response settings.

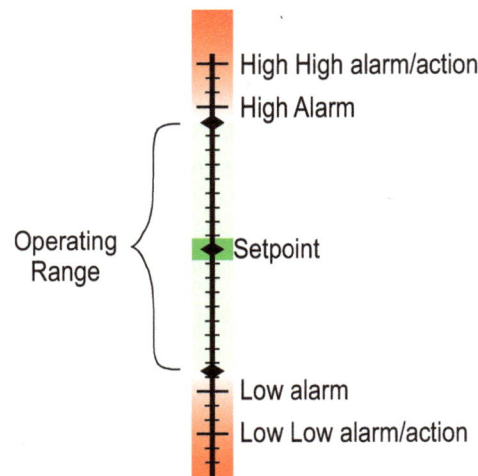

In addition to monitoring equipment operating variables, your work environment may be monitored for the presence of toxic and flammable substances. Environmental emissions may also be monitored. When limits are exceeded, alarms are triggered. In your work area, make sure you know:

- the types of alarm that can be initiated
- the meaning of each alarm
- the response of equipment to each alarm condition (e.g., start/stop fans)
- the potential consequences of each alarm condition

- your required response to the various alarms
- the amount of time you have to respond to each alarm
- when alarms are bypassed for maintenance activities and when the alarms are enabled after maintenance to ensure protection and prevent false alarms being triggered

Instrumentation

A variety of instruments can be used to measure and display the value of variables. Readings are typically displayed on gauges and digital screens. Gauges may be located on the equipment or on a nearby panel.

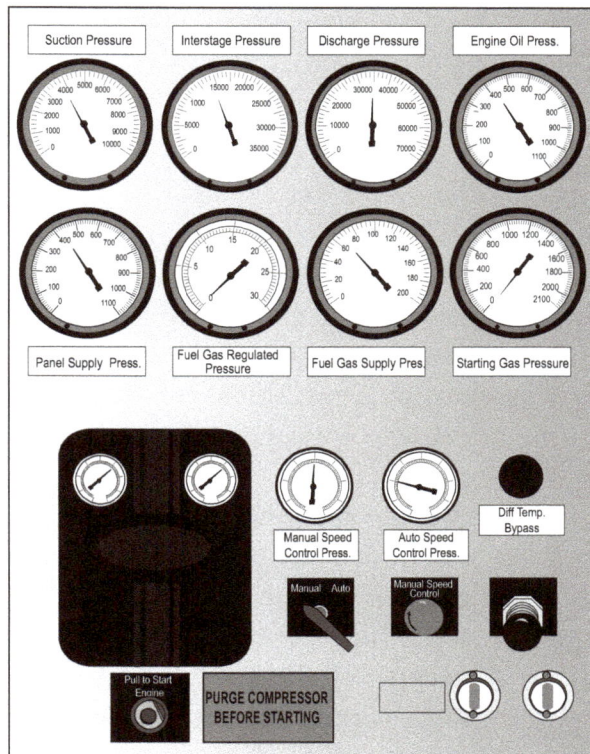

Gauges must be read carefully to obtain an accurate reading, especially when the measurement scale is non-linear.

Digital screens may be located on a local panel, at a central location in the facility (e.g., control room) or at an off-site location. Some equipment may be monitored remotely from portable computers.

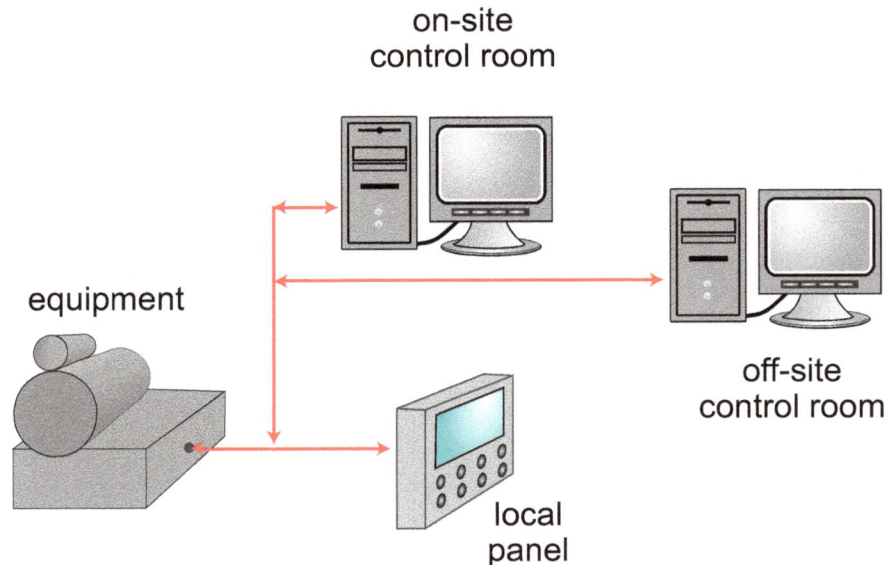

on-site
control room

equipment

off-site
control room

local
panel

Control

Often processing and manufacturing equipment has monitoring and control systems to prevent equipment damage and ensure product quality. The systems monitor critical process and equipment variables. In response to changing variables, the control systems may:

- adjust equipment variables to maintain product quality parameters
- adjust equipment to maintain equipment operating variables (e.g., rpm) within acceptable operating limits
- shut down the equipment if critical equipment variables (e.g., lubrication oil pressure, bearing vibration) are outside acceptable limits
- shut down the system if product variables are outside of specifications (*off spec*)

Control of individual pieces of equipment may be linked so that all the pieces of equipment function as a system. If one piece of equipment fails, there are several possible responses, including:

- a stand-by unit starting up
- throughputs being reduced
- the system shutting down

There are two major process control strategies. A basic control strategy is to start and stop equipment to maintain process variable values within the desired limits. For example, a thermostat in a home will turn a furnace on and off to maintain the desired temperature. If the temperature drops, say by one degree to the lower limit, the furnace turns on. The furnace turns off when the temperature goes above the upper limit (i.e., intermittent operation).

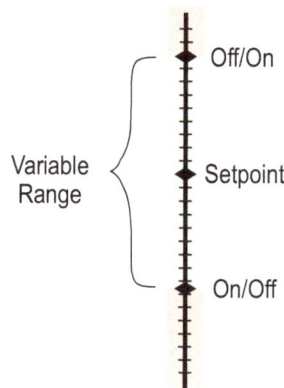

Often the amount of deviation (offset) from the setpoint to the upper and lower limits can be adjusted. If the deviation is very small, the furnace will frequently turn on and off and operate for a short period. If the deviation is very large, the furnace will operate for a longer period of time and the temperature in the room will vary considerably.

Another example of intermittent control is the operation of a sump and pump. A float rises and falls with the level of liquid in the sump. When the float rises to the high level, the pump starts. The pump continues to operate until the float drops to the low level. In this example the sump empties as a batch (i.e., batch control).

NOTE

The *consequence* and/or the cause for a variable to change can be controlled. For example, the liquid level in a tank reaches the high level (consequence). A pump could be started to maintain/reduce the liquid level. (The **consequence** *is being controlled*.) Alternatively, if the tank is being filled by a pump (cause), the pump could be shut down and the liquid would remain at that high level. (The **cause** *is being controlled*.)

A more sophisticated way to control the value of a process variable is to adjust the operation of equipment. For example:
- a valve controlling product flow can be opened and closed slightly to change the flow rate
- an electric motor can be sped up or slowed down to change equipment operating speed

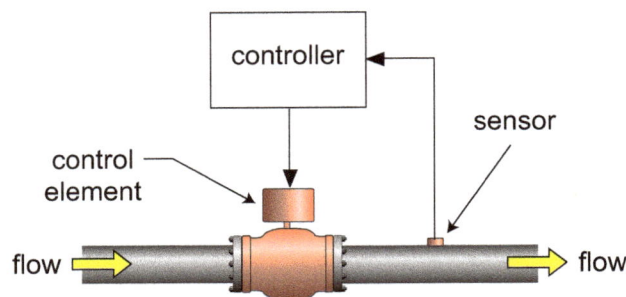

- the sensor sends the value of the measured variable to the controller
- the controller compares the value of the variable to the desired value (setpoint)
- If there is a difference between the measured value and the setpoint, the controller sends a command to the control element to open up or close down on the valve slightly

Process control can be complex. However, control technology has limitations in the ability to accurately maintain setpoint and to respond quickly to changes. There can be a considerable time delay from the time the operator adjusts a setpoint to the time when the control equipment achieves the new value (especially for temperature control).

Often, operators are able to adjust the setpoint. One setpoint adjustment mistake that operators make is to not wait long enough to observe a change occurring. Instead, operators make a larger change to the setpoint to speed up the change process. The process control then overshoots the *desired* setpoint. In response to the overshooting of the *desired* setpoint, the operators make a large change in the setpoint in the opposite direction. The sequence of events repeats itself, causing major swings in the value of the process variable.

To prevent this situation from occurring, operators must:
- make small adjustments to the setpoint, and

- give the automatic controller time to bring the process to the new setpoint before making another setpoint adjustment

When adjusting setpoints, always note the time required for the equipment to respond effectively. To operate effectively under normal and abnormal conditions, you must have a basic knowledge of the control configuration and control strategy for each piece of equipment in the process.

NOTE

When working with stationary equipment that is part of a technical process, the equipment variables may have to be changed to achieve different desired output results or to respond to changes in the input material variables. *MatThink* provides additional information about making changes to equipment to achieve the desired material quality (i.e., output results).

Monitoring

Before startup, during startup, during operation, and during shutdown, use technology and your observations to ensure the equipment:
- can be started and operated
- performs effectively
- achieves the desired results
- does not develop problems that could cause equipment damage or failure
- does not cause harm to PEMEO

Using technology rather than human senses (e.g., hearing, sight) to monitor equipment operation and conditions has several advantages. Technology can often be used to:
- detect variables that cannot be directly sensed by people (e.g., infrared radiation)
- measure variables at limits that cannot be measured by human senses (e.g., high temperatures)
- detect smaller changes in the variables than can be detected by using human senses (e.g., changes in the level of vibration)
- measure variables more accurately under changing conditions than can be measured with human senses

(e.g., during winter and summer, measure the temperature of a pump located outside)

- create detailed long-term records of variable changes over long periods (trending). Without trending, people may not be able to recognize a variable changing very slowly over time (e.g., six months).

Although technology has an important place in monitoring equipment condition and performance, most companies encourage their employees to make on-site observations of the equipment. Employees are encouraged to use their senses of touch, hearing, sight, smell, and, in some cases, taste to assess equipment condition and performance. Using human senses to assess equipment has several benefits:

- A broad range of variables, including the equipment's physical condition, can be monitored, whereas technology used to monitor the equipment may only monitor a few critical variables.
- Some characteristics of a variable may be outside the range of the technology (e.g., low or high vibration frequencies) but can be sensed by the operator.
- Information appearing on remotely located monitors may not be in agreement with local readings (e.g., lubricating oil pressure).
- People who work around the equipment may note changes in performance that are not detected by the monitoring technology (e.g., loose components, leaks, smoothness of operation, a change in the quality of results).

5.5 Exemplary Operating Performance

To be exemplary at operating equipment you must:
- carry out procedures effectively, efficiently, and safely
- have *knowledge* about how the equipment operates and its controls, and how to operate the equipment
- be able to recognize and respond to abnormal operation

To learn how to effectively carry out your company's procedures for operating the equipment, seek answers to the following types of question:
- *What conditions create the need to perform the task?*

- *What are the safety issues?*
- *Who do I communicate with before, during, and after the task is performed?*
- *What are the desired results of performing the task?*
- *What are the acceptable values of the input material variables?*
- *What are the reasons for specific action steps?*
- *Why is the sequence of action steps important?*
- *How do I know I have carried out the action step effectively?*
- *What are the consequences if I do not carry out an action step well?*
- *What are the consequences if the equipment does not respond as expected?*
- *What do I do if the equipment does not respond at all or not as expected?*

To do your job very well, you need knowledge about how equipment operates and how to operate equipment effectively. Focus your learning on the variables that relate to your job. For each variable, determine if it is static or dynamic, controllable or non-controllable. Ask and seek answers to the following types of question:

- *What are the desired output results?*
- *What are the specifications for the input material variables?*
- *What are the input material variables that must be changed by the process?*
- *What are the input material variables that must not be changed?*
- *Can the input materials contain contaminants that can harm the equipment and/or downgrade the desired output results?*
- *How do the materials enter and leave the process?*
- *What equipment variables make the desired change?*
- *What are the technical concepts associated with the variables that cause the change and the reasons the change occurs?*
- *What are the ideal values of the equipment variables that cause the change?*
- *How does the equipment operate?*
- *What are the operating limits of the equipment?*

- *How can I operate the equipment in ways that extend equipment life and condition?*
- *What are the safety and environmental considerations?*
- *How can I operate the equipment in a way that meets the company's goals?*

The following table provides suggestions for focusing your learning about a technical process new to you.

Action Step	Output	Input	Process
Identify outputs, processes, and inputs	• result • product	• raw materials • semi-finished materials • components	• technology • work
Identify critical material variables	• quality • quantity • time • timeliness	• variables that must be changed • variables that must not be changed • contaminants	• method of causing change • variables that change
Identify critical equipment variables	• method of transferring material out of the process	• method of transferring material into the process	• method equipment uses to create the change • equipment operating variables
Determine if variables are static or dynamic	• static • dynamic	• static • dynamic	• static • dynamic
Determine if variables can be controlled	• controllable • non-controllable	• controllable • non-controllable	• controllable • non-controllable
Determine specifications of variables	• target • optimal range • alarm limits • shutdown limits	• target • optimal range • alarm limits • shutdown limits	• target • optimal range • alarm limits • shutdown limits

You also need to have a functional knowledge about abnormal operation. The reasons for equipment to operate in a specific way and the reasons for operating the equipment in a specific way are the basis for recognizing and responding to abnormal operation and conditions.

Both specific equipment and the overall process can operate abnormally. Input and output variables can be outside the desired specifications. Equipment can operate poorly or fail. To respond to abnormal conditions, you need to ask and seek answers to the following types of question:

- *What are the most common abnormal conditions that can occur with the equipment and the materials?*
- *What are the causes for variables to change?*
- *What are the indicators of abnormal conditions?*
- *What are the consequences of abnormal conditions?*
- *How should I respond to the abnormal conditions*
- *What is the amount of time available to respond effectively to abnormal conditions?*

When you have answers to these questions, you will be able to respond more effectively to abnormal operation and minimize the impact on PEMEO.

| NOTE | *MetaThink* provides an in-depth description of reasons, causes, and consequences and their applications to the workplace. |

Having a commitment to developing exemplary operating skills is essential to refining your abilities to work effectively and efficiently. For each specific process, learn the operating and decision-making strategies that contribute to your organization's goals. Receiving formal training, observing others doing the task, and consulting with co-workers all help you learn to work effectively and efficiently.

| IMPORTANT | In technical processes, the following operating objectives are important for the organization and the job: |

- protect people and the environment
- obtain the desired quality of results

- maximize the use of materials
- minimize waste
- minimize energy consumption
- maximize efficiency of work
- minimize equipment wear and damage

Use stationary equipment effectively

This learning activity helps you refine your thinking skills to identify critical output, input, and process variables that:
- define the quality of results
- affect the quality of results
- define the quality of input materials
- create change
- are changed
- can cause harm to equipment and other PEMEO domains
- are affected by PEMEO
- are important to operating equipment effectively

This learning activity also helps you to ask useful *What if... ?* questions. You will also practice identifying the differences in starting and stopping procedures under different conditions and for different reasons.

1. Identify a technical process and answer the following questions. Technical processes involve using stationary equipment that refines materials (e.g., screening gravel for pea size rocks, separating hydrocarbons) or manufactures a product (e.g., windows, plastic parts, pharmaceuticals).

    The technical process is: _____

1a. What are the critical *output* variables (quality, quantity, timeliness)?

    _____

    _____

    _____

1b. What are the critical *input* variables (material properties and characteristics) that have to be changed?

_____

_____

_____

_____

_____

1c. What *input* variables (material properties and characteristics) must **not** be changed by the technical process?

_____

_____

_____

_____

_____

1d. What input variables can change and affect equipment, efficiency of the technical process, and the desired results?

_____

_____

1e. What is/are the *process* variables (e.g., force, heat, pressure, gravity, chemical, electrical) that create the change?

_____

_____

1f. For each variable you listed in question 1e., identify whether the variable is controllable or non-controllable?

_____

_____

_____

1g. Which *input* variables can affect the effectiveness of the process to create the change? How do you adjust the equipment in response to the input variable change?

_____

_____

_____

_____

1h. What foreign material can be in the *input* material that can harm the equipment and/or affect the quality of results? How do you know the foreign material is present?

_____

_____

_____

_____

1i. If foreign material is present, what is the required operator response?

_____

_____

_____

_____

1j. What equipment variables must be monitored?

_____

_____

_____

_____

1k. What are the indicators of poor equipment operation or condition (e.g., noises such as clanging or squealing, changes in temperatures or pressure, varying rpm, leaks at specific locations)? What are the required operator responses?

_____

_____

_____

_____

11.   How can an operator cause premature wear and damage to the equipment (i.e., reduce the life and condition of the equipment)?

_____

_____

_____

_____

2.   For most technical processes, there is always a possibility of harm being caused to PEMEO. Fill out the following table. For each category of PEMEO, identify two conditions, actions, and/or events that could cause harm. Identify the potential consequences.

Technical process: _____

Potential Harm to PEMEO		
Category	Condition, action, event	Consequences
People		
Equipment		
Materials (including structures)		
Environment		
Organization		

3. Using the technical process identified in question 1, what alarms can be initiated and what are the required operator responses?

_____

_____

_____

_____

4. One strategy that people often use to learn about similar equipment and tasks is to look for differences. First, people learn about a specific technology or task. Then, when working with technology or a task that are only slightly different, they look for differences. This thinking strategy makes learning and working effective and efficient. Using the technical process identified in question 1, complete the following table to explain the differences between different types of shutdown.

Technical process: _____

Shutdowns	Differences in Operator Actions	Differences in Equipment Responses
Differences between a normal and emergency shutdown		
Differences between a normal and maintenance shutdown		

5. Using the technical process identified in question 1, identify the differences of the pre-start checks and startup after a normal shutdown *and* the pre-start checks and startup after:

- a maintenance shutdown
- an emergency shutdown

Technical process: _____

	Differences in Pre-Start Checks	Differences in Startup Actions
Differences of the pre-start checks and startup after a normal shutdown *and* the pre-start checks and startup after a maintenance shutdown		
Differences of the pre-start checks and startup after a normal shutdown *and* the pre-start checks and startup after an emergency shutdown		
Differences of the normal pre-start checks and startup in hot weather *and* the pre-start checks and startup in cold weather		

6. Use the technical process listed in question 1. In the following table, list three critical variables you pay special attention to when performing each step of the task.

Technical process: _____

Actions	Critical Variables
Shut down equipment	• • •
Plan the work	• • •

(continued)

Actions	Critical Variables
Perform pre-start checks	• • •
Start equipment	• • •
Operate equipment	• • •

7. Using the technical process identified in question 1, what are the communication requirements for each of the following actions?

Technical Process: _____

Actions	Communication Requirements
Entering the area	
Before starting equipment	
Before adjusting equipment	
Before shutting down equipment	
After shutting down equipment	

8. In the Job Aid there are two lists that apply to stationary equipment:
   • *Critical Thinking Questions about Technical Processes Using Stationary Equipment*
   • *Critical Thinking Questions about Operating Stationary Equipment*

Check off the questions that apply to your work and equipment. Add any questions that are work or equipment-specific.

Equipment Life and Reliability

Reliable equipment starts when needed and operates effectively and efficiently throughout the operation. Equipment reliability is affected by both operating and maintenance activities. For example, servicing vehicles before winter helps ensure they will start and keep operating in cold weather.

When equipment is reliable, it is usually available for production. However, maintenance activities can restrict the availability of equipment. Maintenance must be planned so that, when a piece of equipment is taken out of service, there is minimal effect on production.

Equipment reliability and equipment life are interrelated; activities that contribute to keeping equipment reliable usually contribute to extending equipment life.

6.1 Equipment Life

There are three common definitions for equipment life—safe life, rated life, and maximum life.

Safe Life

Safe life refers to the amount of safe service the equipment or its components can provide. Should a component fail, there could be catastrophic consequences for safety, the environment, equipment, and/or process.

Length of service and total use are two criteria often used to determine safe life. When the safe life limit has been met, the equipment or components are replaced to ensure reliability (and, in many cases, safety). For example, some aircraft components must be replaced after a specific amount of flying time (e.g., 1000 flying hours); other components are replaced after a specific number of uses (e.g., 5000 landings).

Rated Life

Manufacturers may state the expected life of equipment (e.g. operating hours, distance travelled) before major repairs or equipment replacement are required. Should the equipment or components fail, production may be hampered and repair costs could be significant. However, safety and the environment are usually not at risk.

In some cases, critical equipment components dictate the rated life of the equipment. For example:
- The drum in a photocopier has to be replaced after a specified number of copies has been made. If the drum is not replaced as specified, the quality of the copies will decrease.
- The timing belt in some vehicle engines should be replaced after 100,000 km. If the belt fails, the engine will be severely damaged, requiring major repairs.

Manufacturers may back up their claims by offering warranties. For a warranty to be valid, equipment must be used for the stated applications, and operated and maintained in accordance with the manufacturer's specifications.

Maximum Life

An organization's operating philosophy affects equipment life. Depending on the circumstances, an organization may choose to operate equipment until it fails or may choose to operate equipment for reliability and extended life. To get the most return on the capital investment in equipment, organizations often make efforts to extend the life of equipment. Several factors affect the rate at which equipment deteriorates, including:

- equipment application (e.g., a vehicle that continually starts and stops in heavy traffic on city streets wears faster than a vehicle driven on paved highways that does not start and stop as frequently)
- manner in which the equipment is operated (e.g., speed when going over potholes)
- quality and frequency of maintenance activities

Determining the end of equipment life can be a complex process and may involve assessing several factors, including:

- reliability
- safety
- efficiency of operation
- effectiveness of operation
- cost of immediate and future repairs versus cost to replace the equipment
- cost of downtime required to repair the equipment
- period of time required to procure and replace the equipment
- period of time the equipment will be needed (e.g., a specific project may require equipment that is not normally used or a line of business may be shut down)

6.2 Equipment Reliability

As mentioned above, reliable equipment starts and operates effectively and efficiently throughout the operation. As equipment wears, performance may diminish. A drop in performance can have adverse effects on PEMEO. For example:

- peoples' safety may be compromised (e.g., worn brakes)
- employees may have to work harder to achieve the desired results (e.g., dull cutting tools)
- other equipment and facilities could get damaged (e.g., broken metal components from upstream of a pulp refiner could damage the refiner)

- quality of products or work results may decrease and waste increase (e.g., worn sanders in a furniture production line result in some parts being rejected)
- energy consumption and pollution may increase (e.g., poorly operating internal combustion engine)
- damage to the environment may occur (e.g., oil pipeline valve leaks)

Some components and equipment perform critical functions. A function is considered critical if it has a major effect on PEMEO. Fail-safe and backup strategies may be applied to critical components and equipment such as:

- equipment which performs critical operating functions. In computer control centers, if the host computer fails, a stand-by computer takes over.
- equipment and systems that could be a safety hazard in the event the energy source fails. At facilities such as hospitals and gas plants, if the source of electrical energy fails, a temporary electrical backup system activates to supply electricity to critical systems (e.g., fire detection, emergency lighting, and life support systems).
- safety devices which monitor the work environment in the event the energy source fails. Toxic or flammable gas detectors have an electrical backup system that activates in the event of a power failure.
- critical equipment fails in the safest position. On gas process systems, critical valves either fail closed or open if the source of energy is lost.

NOTE

Sometimes duplicate components and equipment which perform critical functions are provided to ensure the system continues to operate should the equipment fail or be taken out of service for maintenance. For example, aircraft engines have twin ignition systems and pipeline pumping stations often have a spare pump.

6.3 Maintain Reliability and Extend Equipment Life

This section describes strategies used to maintain reliability and extend equipment life:

- operational inspections
- major inspections
- reactive maintenance
- preventive maintenance
- predictive maintenance
- reliability-centered maintenance
- troubleshooting

Operational Inspections

Equipment is checked before, during, and immediately after equipment operation. Some organizations use a checksheet which operators routinely complete, distribute, and file. If there is a change in equipment condition or function, action may be taken to correct the problem. In some cases, if poor performance of one component is not corrected, the component could fail, causing other components to be damaged or fail. For example, if a thrust bearing on a steam turbine fails, the rotor could shift, causing the rotor blades to collide with the stator blades. The collision results in catastrophic equipment failure.

Major Inspections

With use, most equipment wears, decreasing performance and reliability and increasing the risk to PEMEO. To inspect critical components, equipment often must be shut down and opened or dismantled for inspection. Unfortunately, up to seventy percent of equipment component damage occurs during these inspections due to problems during the disassembly and/or assembly—rust and scale makes it difficult to remove components without causing damage and poor maintenance practices can damage machined surfaces.

In manufacturing and process industries, often one piece of equipment cannot be shut down for inspection without shutting down the entire system. These industries carry out equipment inspections, refurbishing, and retrofitting by scheduling a shutdown of a subsystem or entire plant. Scheduled plant shutdowns are often called turnarounds. During a turnaround, a variety of inspection techniques may be used to ensure the integrity of the equipment. Methods of inspection include:

- visual
- borescope inspection cameras
- x-ray
- gamma ray
- magnaflux
- pressure test

Identified problems may be corrected at the time or at some future turnaround. There is a financial trade-off between correcting the problem immediately or later. Fixing the problems as part of the turnaround may require more time and considerable unanticipated costs. Extending the turnaround time also causes additional losses of productivity and revenue. Operating without correcting the problem may require additional operating inspections and/or operating the equipment at a lower throughput. Should the equipment fail before the next turnaround, there will be unanticipated loss of production, extra costs to make repairs, and the potential to cause customer dissatisfaction.

Setting up for the turnaround can cause considerable disruption of the job site. Scaffolding may have to be erected and new equipment stored. Additional personnel brought in to prepare the site and plan the turnaround can increase the safety risks; having more people on site adds to the complexity of the work environment and the new workers may not be familiar with the site. Activities must be carefully planned, coordinated, and supervised.

Specific turnaround procedures are needed for shutting down a plant, preparing equipment for maintenance, and starting up a plant to minimize the risk to PEMEO. Sometimes new procedures must be developed because the plant conditions and hazards can change from one turnaround to the next.

Reactive Maintenance

Reactive maintenance often involves responding to a request for maintenance. These requests can be made for many reasons, including:

- abnormal equipment operation
- unsatisfactory equipment performance
- equipment failure
- need for equipment modifications
- required change to the equipment's fixed variables
- need to upgrade equipment

In large organizations, maintenance requests are usually documented via paper or electronic work orders. High priority requests may first be placed orally, followed by a written work order. Reactive maintenance work is often ranked in priority based on criteria such as:

- safety of personnel
- protection of the environment
- sustainability of production
- simplification of work
- cost savings

In some instances, a minimal investment in maintenance is made to sustain equipment life:

- production takes priority over equipment condition
- the equipment will be replaced upon failure
- the equipment or facility will be permanently shut down
- the facility is to be sold

Jerry rigging

Jerry rigging is an unsatisfactory fix to temporarily make do. For example:

- A maintenance person did not have the correct vibration probe. He installed a vibration probe that resulted in the vibration reading being only half of what it should have been. He told operators to double the reading they observed on the vibration panel. The vibration would have to be twice as severe as normal to trigger the alarm and shut down the equipment. The bearing could vibrate excessively and eventually fail.

- The wood on an extension ladder was beginning to split. The split was wrapped with wire to make do.

There are several concerns about jerry rigging. The solution:
- may not be reliable
- may end up being a long-term solution
- could become an acceptable way to fix equipment
- could increase the potential for further damage (e.g., a failed pump bearing could cause other pump components to be severely damaged)

Documenting equipment condition and failure

If there is poor documentation and communication of events, especially between work shifts, the frequency of the problems may not be noted. Frequent failure of a specific component and frequent need for a specific reactive maintenance task should be investigated; ways to eliminate the problem must be determined.

Some companies have comprehensive maintenance databases. These databases can be examined to determine:
- the frequency of specific reactive maintenance tasks
- the maintenance history of specific equipment
- quantity of specific components used for maintenance

This information is valuable in determining reoccurring problems and possible solutions. A thorough diagnostic process may have to be used to determine causes for failures and possible solutions. Developing maintenance procedures can also improve the effectiveness and efficiency of the maintenance work.

Preventive Maintenance

With use, equipment conditions change. These changes can adversely affect equipment performance, reliability, and life. The purpose of preventive maintenance is to minimize the adverse effects by carrying out scheduled routine maintenance such as:
- changing engine lubricating oil
- replacing process filters
- replacing ignition components

- calibrating monitoring equipment
- overhauling equipment
- routine cleaning of equipment components
- maintaining machinery not in use that have rotating parts

Equipment condition determines the schedule for routine maintenance. However, when specific conditions are not known, time may be used as the determining factor to schedule routine maintenance. Here are examples:

Examples of Factors used to Schedule Routine Maintenance	
Routine maintenance	Scheduling factors
Change engine lubricating oil	kilometers, oil color, operating hours
Replace process filters	differential pressure across filter, operating hours
Replace engine ignition components	decreased equipment performance, operating hours
Overhaul equipment	decreased equipment performance, operating hours

Manufacturers recommend types and schedules of routine maintenance and specifications for replacement components. These recommendations may be based on typical equipment application and operating environment and may not be suitable for a specific situation. For example:
- If time is the determining factor for lube oil changes, the schedule may be too frequent for clean operating environments or not frequent enough for dirty operating environments.
- The recommended oil may not function adequately if the equipment continuously operates under heavy load or if the temperature of the product and operating environment is higher than normally expected.

To establish maintenance schedules and component replacement specifications, a company may carry out tests such as oil analysis and failed components analysis. The company may also use a computer database for maintenance

management. The database may provide information on
- frequency of equipment failure
- routine maintenance schedules
- work scheduled
- time required to carry out maintenance tasks
- labor costs
- component inventory and prices
- suppliers
- parts on order

Companies that keep comprehensive records may reassess the routine maintenance schedules as a result of a:
- history of frequent component failures and poor equipment performance
- commitment to minimize maintenance costs
- commitment to maximize equipment availability, performance, reliability, and life

Conducting equipment inspections and routine maintenance can potentially downgrade the equipment. For a given type of technology or task, some specific procedures and precautions must be taken to prevent downgrading. For example:
- A worker repairing electronic hardware must be electrically grounded to prevent static electricity from damaging components.
- Bare hands must not touch halogen lamps and other optical components. Oils from the hands can contaminate the surface and affect performance and/or shorten component life.
- When servicing a gasoline internal combustion engine, dirt must not fall into the engine's combustion cylinders and lubricating oil sump to prevent component damage.

Written procedures should contain action steps and precautions to prevent downgrading equipment or contaminating materials.

Machinery not in use that has rotating parts can deteriorate—shafts sag and bearings become pitted. To prevent sagging of shafts, the shafts are rotated $1\frac{1}{4}$ turns on a scheduled basis (e.g., bi-weekly). Bearings are treated with a preservative recommended by the manufacturer to prevent pitting.

Predictive Maintenance

Predictive maintenance involves monitoring the "health" of critical components to predict when the components will fail. Predicting the time of component failure has several advantages, including:

- performing maintenance based on component condition rather than on time
- further ensuring equipment reliability
- preventing a component from failing and causing damage to other components
- providing advanced notice of maintenance so that work can be scheduled to minimize production and use maintenance resources effectively

Vibration and temperature are two variables that can be monitored to determine equipment health. In a large facility, thousands of components could be monitored. However, for cost effectiveness, only critical components are monitored for vibration and/or temperature. A company starting a predictive maintenance program uses several factors to identify equipment and components to be monitored, including the following factors:

- critical equipment affecting production, safety, and environment
- critical components affecting equipment health and cost of repairs
- frequently failing components

The predictive maintenance process involves:

- monitoring variables of the critical components in accordance with a schedule (e.g., once per day at a specified time or event)
- accurately recording the data for each component
- relating the changes in variables to other changes (e.g., change in material characteristics, change in throughputs)
- noting changes in the variables as the components wear (trending)*
- predicting component failures
- deciding on the best course of action

   * data may be presented numerically or graphically on a computer terminal or printout

Reliability-Centered Maintenance

Budgets restrict carrying out as much maintenance as is desired. To make the most use of limited budgets to prevent production disruptions, critical equipment that can shut down production receives the most maintenance. Reliability-centered maintenance involves identifying the critical equipment, determining the most likely cause for failure, and carrying out preventive maintenance.

Troubleshooting

Troubleshooting equipment to determine the cause for abnormal operation or failure can be difficult, especially if there is more than one source causing the problem. Identifying the actual causes of the problem is important because misdiagnosis can be very costly. Replacing the wrong component is costly and does not solve the problem.

Equipment problems often occur as a result of a sequence of events. There can be an immediate cause and a root cause (originating cause) for a problem. A root cause leads to an immediate cause for equipment problems. Sometimes treating the immediate cause does not minimize the potential for the problem to recur (not a long-term solution).

Determining the root cause is important to avoid repeated costly and ineffective corrective actions. For example, a thrust bearing on a steam turbine shaft fails, causing the turbine rotor to collide with the stator. There is no assurance that replacing the failed bearing will solve the problem. The root cause for the bearing failure must be determined. In this example, a number of reasons for the bearing failure might be considered, including a manufacturer defect, use of the wrong lubricant, inadequate cooling, or improper operating procedures.

Accurate, current information makes the troubleshooting process more effective and efficient. Information useful for troubleshooting includes:
- operating records
- maintenance records
- equipment history

- material specifications
- technical drawings
- equipment specifications
- troubleshooting charts
- data from diagnostic instruments
- employee experiences and observations

The quality of the maintenance affects the reliability and life of equipment:

- quality work is essential
- parts are often replaced in kind with parts produced by the manufacturer
- a third party manufacturer may be used because their parts work better in unfavorable operating conditions (e.g., excessive heat or dust)
- a third party manufacturer may be used because their parts cost much less but unfortunately are also less reliable

LEARNING ACTIVITY 4

Equipment reliability and life

This learning activity helps you to refine your thinking skills to identify how poorly performing equipment can affect PEMEO. You will also identify the required maintenance of a piece of equipment to ensure the equipment's reliability and to extend equipment life.

1. Select a piece of equipment and determine the adverse effects on PEMEO should the equipment perform poorly or fail. Complete the following table.

    The equipment is: _____

	Adverse Effect	
Category	Equipment performs poorly	Equipment fails
People		
Equipment		
Materials (including structures)		
Environment		
Organization		

2. In the following table, identify in your company three pieces of equipment and the maintenance requirements.

Equipment	Types of Maintenance	Maintenance Requirements
	☐ reactive ☐ preventive ☐ predictive ☐ reliability-centered	
	☐ reactive ☐ preventive ☐ predictive ☐ reliability-centered	
	☐ reactive ☐ preventive ☐ predictive ☐ reliability-centered	

Job Aid

Using this Job Aid

Identifying the critical variables for the equipment you operate and maintain:

- helps you to focus your efforts to work more effectively and efficiently.
- provides valuable information when mentoring others to operate and maintain the equipment.

Instructions

1. Select a piece of equipment you operate and identify the task application.

2. For each set of questions, check off the questions that are most important to operating the equipment effectively and efficiently to achieve the desired results.

3. For the variables you have identified, make notes explaining the specific variables and their setpoints or ranges.

4. Explain what can go wrong and how to respond.

EquipThink™

Equipment:_____

Task application: _____

_____

_____

_____

_____

_____

_____

_____

_____

_____

_____

_____

_____

_____

_____

_____

_____

_____

_____

_____

_____

_____

_____

_____

_____

_____

_____

Generic Questions for Using Tools and Equipment Effectively and Efficiently

Output Variables

- ☐ What *output* variables define quality of results?
- ☐ How are the *output* variables measured?
- ☐ What are the targets or optimal ranges for *output* variables?
- ☐ What is the impact of *output* materials on PEMEO?
- ☐ What is the impact of PEMEO on *output* materials?

Input Variables

- ☐ What are the *input* variables?
- ☐ What *input* variables are changed by the equipment?
- ☐ What *input* variables are static? Dynamic?
- ☐ How are the *input* variables measured?
- ☐ What are the targets or optimal ranges for *input* variables?
- ☐ What *input* variables are controllable? Non-controllable?
- ☐ What *input* variables must **not** be changed?
- ☐ What is the impact of PEMEO on *input* materials?
- ☐ What is the impact of *input* materials on PEMEO?

Work Process Variables

- ☐ What *equipment* variables change the materials?
- ☐ How are the *equipment* variables that change the materials measured?
- ☐ What *equipment* variables that change the materials are static? Dynamic?
- ☐ What *equipment* variables that change the materials are controllable? Non-controllable?
- ☐ What *equipment* variables can cause damage to the desired results?
- ☐ What work processes (skill) can cause damage to the materials?
- ☐ What can cause damage and wear to the equipment?
- ☐ How can the work processes and equipment cause damage to adjacent components, equipment, and fixtures?
- ☐ What are the safety issues?
- ☐ How can the environment be affected?
- ☐ What work strategies contribute to the effectiveness of the work?
- ☐ What work strategies contribute to the efficiency of doing the work?
- ☐ What organizational factors affect the work and can have adverse effects on PEMEO?

(continued)

EquipThink™

Notes

Equipment:_____

Task application: _____

_____

_____

_____

_____

_____

_____

_____

_____

_____

_____

_____

_____

_____

_____

_____

_____

_____

_____

_____

_____

_____

_____

_____

Generic Questions for Using Tools and Equipment Effectively and Efficiently

Organizational conditions, actions, or events that adversely affect the job and work group

- ☐ Are there changes in the work process?
- ☐ Is there a shortage of staff?
- ☐ Is there a lack of competent workers assigned to a specific task?
- ☐ Has a work group introduced new hazards?
- ☐ Has a third party failed to maintain equipment/facilities?
- ☐ Has the supplier changed standards or composition of components or materials?
- ☐ Is there an unpredicted large customer order?
- ☐ Is there a cancellation or delay of a large customer order (e.g., may have excessive inventory on site)?
- ☐ Is there a failure to communicate priorities?
- ☐ Is there a lack of documentation?
- ☐ Is the administrative process inadequate for maintaining inventory?
- ☐ Is there a failure to carry out routine safety or quality inspections?
- ☐ Is there a failure to follow up on identified safety or quality deficiencies?

Critical Thinking Questions about Work Processes Using Mobile Equipment

Output Variables

- ☐ What *output* variables define quality of results?
- ☐ How are the *output* variables measured?
- ☐ What are the targets or optimal ranges for *output* variables?
- ☐ What is the impact of *output* materials on PEMEO?

Input Variables

- ☐ What material *input* variables must be changed?
- ☐ How are the *input* variables measured?
- ☐ What are the material *properties* and *characteristics* that can affect the results?
- ☐ What are the material *properties* and *characteristics* that can cause harm to the equipment?
- ☐ What *input* variables are static? Dynamic?
- ☐ Can the input material *properties* and *characteristics* change?
- ☐ How can a change in input material *properties* and *characteristics* affect the quality of results?
- ☐ Can a change in input material *properties* and *characteristics* affect the work strategy?

(continued)

Critical Thinking Questions about Work Processes Using Mobile Equipment

Input Variables *(continued)*

☐   What *input* variables must **not** be changed?

☐   What surrounding conditions can affect the methods of work and safety?

☐   What is the impact of PEMEO on *input* materials?

☐   What is the impact of *input* materials on PEMEO?

Notes

Equipment:_____

Task application: _____

_____

_____

_____

_____

_____

_____

_____

_____

_____

_____

_____

_____

_____

_____

_____

_____

_____

_____

_____

_____

_____

Critical Thinking Questions about Work Processes Using Mobile Equipment

Work Process Variables

- ☐ Are there changes in the work process?
- ☐ What *equipment* variables change the materials or achieve the desired results?
- ☐ What *equipment* variables that change the materials are static? Dynamic?
- ☐ What *equipment* variables that change the materials are controllable? Non-controllable?
- ☐ How does the equipment work?
- ☐ What are the capabilities of the equipment (e.g., size)?
- ☐ What equipment variables can cause damage to the desired results?
- ☐ What work *processes* (skill) can cause damage to the materials?
- ☐ What can cause damage and wear to the equipment?
- ☐ How can the work *processes* and equipment cause damage to adjacent components, equipment, and fixtures?
- ☐ What are the safety issues?
- ☐ How can the environment be affected?
- ☐ What work strategies contribute to the effectiveness of the work?
- ☐ What work strategies contribute to the efficiency of doing the work?
- ☐ What organizational factors affect the work and can have adverse effects on PEMEO?

Critical Thinking Questions about Operating Mobile Equipment

Park Equipment

- ☐ Do I refuel before parking the equipment?
- ☐ Is there a designated area and equipment orientation for parking?
- ☐ Where do I park equipment to prevent interfering with other work and possible damage to the equipment?
- ☐ How do I prevent the equipment from rolling?
- ☐ How do I position operating components to prevent harm to people?
- ☐ How do I position operating components to prevent damage from the elements?

Stop Equipment

- ☐ Do I have to let the equipment cool down before shutting it down?
- ☐ How do I perform a normal shutdown?
- ☐ How do I perform a maintenance shutdown?
- ☐ How do I perform an emergency shutdown?
- ☐ How do I secure the equipment and load to prevent theft?
- ☐ What documentation is required?
- ☐ Who do I report to about equipment condition and damage?

(continued)

Equipment:_____

Task application: _____

_____

_____

_____

_____

_____

_____

_____

_____

_____

_____

_____

_____

_____

_____

_____

_____

_____

_____

_____

_____

_____

_____

_____

_____

_____

Critical Thinking Questions about Operating Mobile Equipment

Plan the Work

- ☐ What are the *output* and *input* variables?
- ☐ What type of equipment will be used?
- ☐ What is the capacity of the equipment?
- ☐ What surrounding conditions can affect safety and the work?
- ☐ Will weather conditions affect the work and safety?
- ☐ What changes to materials and the surroundings could occur while working?
- ☐ For the given task, surroundings, materials, and equipment, what work strategy will be most effective, efficient, and safe?
- ☐ What communication and documentation are required before, during, and after working?

Perform Pre-work Checks

- ☐ Where do I look for damage?
- ☐ What are the equipment operational checks (e.g., fluid levels)?
- ☐ What adjustments are required to operate the equipment comfortably and safely (e.g., seat, mirrors)?
- ☐ Are there people or materials that could be harmed upon startup?
- ☐ Are there tools and materials that could cause harm upon startup?
- ☐ What variables should I monitor upon startup?
- ☐ What noises, vibration, and movement of parts indicate abnormal operation?
- ☐ What safety devices should I check for operation (e.g., lights, brakes)
- ☐ What moving components should I check for operation?

Operate the Equipment

- ☐ What are the desired results?
- ☐ What are the safety issues in operating the equipment?
- ☐ Can people enter the work area and be struck by the equipment or materials?
- ☐ How can I damage surrounding equipment and structures?
- ☐ What can I do to prevent damaging surrounding equipment and structures?
- ☐ What are the indications that the equipment is operating abnormally?
- ☐ What are the criteria for safe, effective, and efficient operating skills?
- ☐ What changes in operating strategies and skills are required if the input material changes?
- ☐ What operating actions can cause stress on the equipment and reduce equipment life?
- ☐ What communication is required while operating?

EquipThink™

Notes

Equipment:_____

Task application: _____

_____

_____

_____

_____

_____

_____

_____

_____

_____

_____

_____

_____

_____

_____

_____

_____

_____

_____

_____

_____

_____

_____

_____

_____

Critical Thinking Questions about Technical Processes Using Stationary Equipment

Output Variables

☐ What *output* variables define quality of results?

☐ How are the *output* variables measured?

☐ What are the targets or optimal ranges for *output* variables?

☐ Can the desired quality of results change?

☐ What is the impact of the *outputs* on PEMEO?

☐ What is the impact of PEMEO on the *outputs*?

Input Variables

☐ What are the *input* variables?

☐ Which *input* variables are changed by the process?

☐ Which *input* variable must not be changed?

☐ Which *input* variables are static? Dynamic?

☐ Which *input* variables are controllable? Non-controllable?

☐ How are the *input* variables measured?

☐ What are the targets or optimal ranges for *input* variables?

☐ Can the *input* materials contain contaminants that could affect the equipment and desired results?

☐ What is the impact of *input* materials on PEMEO?

☐ What is the impact of PEMEO on *input* materials?

Process Variables

☐ What *process* variables change the materials?

☐ How are the *process* variables that change the materials measured?

☐ Which *process* variables that change the materials are static? Dynamic?

☐ Which *process* variables that change the materials are controllable? Non-controllable?

☐ Which *process* variables can downgrade the results?

☐ What is the impact of *process variables* on PEMEO?

☐ What is the impact of PEMEO on the *process*?

Change in Variables

☐ What are the indications that the *process* variable has changed?

☐ What are the potential consequences if a *process* variable changes?

☐ What do I do if the *process* variables change?

☐ How much time do I have to respond effectively to a change in a variable?

☐ If the *input* variables change, what changes do I have to make to the *process* variables?

☐ If the quality standards for results change, what changes to the *input* or *process* variables do I have to make?

EquipThink™

Equipment:_____

Task application: _____

_____

_____

_____

_____

_____

_____

_____

_____

_____

_____

_____

_____

_____

_____

_____

_____

_____

_____

_____

_____

_____

_____

_____

_____

Critical Thinking Questions about Operating Stationary Equipment

Stop Equipment

☐ Whom do I communicate with about the pending shutdown?

☐ Do I have to let the equipment cool down before stopping the equipment?

☐ How do I perform a normal shutdown?

☐ How do I perform a maintenance shutdown?

☐ How do I perform an emergency shutdown?

☐ How do I secure the equipment and load to prevent injury?

☐ What documentation is required?

☐ Whom do I report to about equipment condition and damage?

Perform Pre-start Checks

☐ Do I have to communicate with someone before I enter the area?

☐ Where do I look for damage?

☐ Are the pre-start checks different as a result of different shutdowns?

☐ Does the weather affect the types of pre-start checks performed?

☐ What operational checks (e.g., fluid levels) do I perform?

☐ What adjustments are required to start the equipment?

☐ Are there people or materials that could be harmed upon startup?

☐ Are there tools and materials that could cause harm upon startup?

☐ What variables should I monitor upon startup?

☐ What noises, vibration, and movement of parts indicate abnormal operation?

☐ What safety devices should I check for operation (e.g., lights, brakes)

☐ What moving components should I check for operation?

Start Stationary Equipment

☐ Whom do I inform about the pending startup?

☐ Are the startup procedures different as a result of the type of shutdown?

☐ Are the startup procedures different for summer and winter?

☐ What are the reasons for specific startup steps?

☐ Why is the sequence of the startup steps important?

☐ How do I start the equipment?

☐ What are the consequences if the equipment does not respond at all or as expected?

☐ What variables should I monitor upon startup?

☐ What noises, vibration, and movement of parts indicate abnormal operation?

☐ What safety devices should I check for operation?

☐ What moving components should I check for operation?

(continued)

EquipThink™

Notes

Equipment:_____

Task application: _____

_____

_____

_____

_____

_____

_____

_____

_____

_____

_____

_____

_____

_____

_____

_____

_____

_____

_____

_____

_____

_____

_____

_____

_____

_____

_____

Critical Thinking Questions about Operating Stationary Equipment

Adjust the Equipment

- ☐ What are the desired results?
- ☐ What are the safety issues in operating the equipment?
- ☐ Can people enter the work area and be contacted by the equipment and materials?
- ☐ Which process and equipment variables should I routinely monitor?
- ☐ What operating actions can cause stress on the equipment and reduce equipment life?
- ☐ How can I operate the equipment in ways that extend equipment life and condition?
- ☐ How can changes in equipment and the process affect upstream and downstream operation?
- ☐ What changes do I have to make to the process variables if the input variables change?
- ☐ What changes to the input or process variables do I have to make if the desired results change (e.g. quality, quantity)?
- ☐ Whom do I communicate with before adjusting equipment and processes?
- ☐ What types of abnormal process operation can occur?
- ☐ What types of abnormal equipment operation can occur?
- ☐ What are the indicators of abnormal process or equipment operation?
- ☐ What are the potential consequences of abnormal operation?
- ☐ What action do I take in response to abnormal operation?
- ☐ How much time do I have to respond effectively to abnormal operation?
- ☐ How can the process and equipment affect PEMEO?